Advances in Experimental Medicine and Biology

Volume 1284

Advances in Experimental Medicine and Biology provides a platform for scientific contributions in the main disciplines of the biomedicine and the life sciences. This series publishes thematic volumes on contemporary research in the areas of microbiology, immunology, neurosciences, biochemistry, biomedical engineering, genetics, physiology, and cancer research. Covering emerging topics and techniques in basic and clinical science, it brings together clinicians and researchers from various fields.

Advances in Experimental Medicine and Biology has been publishing exceptional works in the field for over 40 years, and is indexed in SCOPUS, Medline (PubMed), Journal Citation Reports/Science Edition, Science Citation Index Expanded (SciSearch, Web of Science), EMBASE, BIOSIS, Reaxys, EMBiology, the Chemical Abstracts Service (CAS), and Pathway Studio.

2019 Impact Factor: 2.450 5 Year Impact Factor: 2.324

More information about this series at http://www.springer.com/series/5584

Hao Wang
Editor

Neural Circuits of Innate Behaviors

Springer

Editor
Hao Wang
Department of Neurobiology, Key Laboratory for
Biomedical Engineering of Education Ministry
Zhejiang University School of Medicine
Hangzhou, Zhejiang, China

ISSN 0065-2598 ISSN 2214-8019 (electronic)
Advances in Experimental Medicine and Biology
ISBN 978-981-15-7085-8 ISBN 978-981-15-7086-5 (eBook)
https://doi.org/10.1007/978-981-15-7086-5

This Springer imprint is published by the registered company Springer Nature Singapore Pte Ltd.
The registered company address is: 152 Beach Road, #21-01/04 Gateway East, Singapore 189721, Singapore

Contents

Contributors

Xiaowei Chen Brain Research Center and State Key Laboratory of Trauma, Burns, and Combined Injury, Third Military Medical University, Chongqing, China

Sijia Hao Department of Neurobiology, Key Laboratory for Biomedical Engineering of Education Ministry, Zhejiang University School of Medicine, Hangzhou, Zhejiang, China

Department of Neurosurgery of Second Affiliated Hospital, Key Laboratory for Biomedical Engineering of Education Ministry, Zhejiang University School of Medicine, Hangzhou, Zhejiang, China

Mohamed Helmy Department of Neurobiology, Key Laboratory for Biomedical Engineering of Education Ministry, Zhejiang University School of Medicine, Hangzhou, Zhejiang, China

Department of Neurosurgery of Second Affiliated Hospital, Key Laboratory for Biomedical Engineering of Education Ministry, Zhejiang University School of Medicine, Hangzhou, Zhejiang, China

Wenjun Jin Brain Research Center and State Key Laboratory of Trauma, Burns, and Combined Injury, Third Military Medical University, Chongqing, China

Xing-Yu Li Institute of Neuroscience, State Key Laboratory of Neuroscience, CAS Center for Excellence in Brain Science and Intelligence Technology, Chinese Academy of Sciences, Shanghai, China

University of the Chinese Academy of Sciences, Beijing, China

Han Qin Brain Research Center and State Key Laboratory of Trauma, Burns, and Combined Injury, Third Military Medical University, Chongqing, China

Chaoran Ren Guangdong-Hongkong-Macau Institute of CNS Regeneration, Ministry of Education CNS Regeneration Collaborative Joint Laboratory, Jinan University, Guangzhou, China

Qian Tao School of Basic Medicine, Jinan University, Guangzhou, China

Hao Wang Department of Neurobiology, Key Laboratory for Biomedical Engineering of Education Ministry, Zhejiang University School of Medicine, Hangzhou, Zhejiang, China

Department of Neurosurgery of Second Affiliated Hospital, Key Laboratory for Biomedical Engineering of Education Ministry, Zhejiang University School of Medicine, Hangzhou, Zhejiang, China

NHC and CAMS Key Laboratory of Medical Neurobiology, MOE Frontier Science Center for Brain Research and Brain Machine Integration, School of Brain Science and Brain Medicine, Zhejiang University, Hangzhou, Zhejiang, China

Lieju Wang Department of Anesthesiology, The Second Affiliated Hospital, School of Medicine, Zhejiang University, Hangzhou, Zhejiang, China

Ying Wu Department of Anesthesiology, The Second Affiliated Hospital, School of Medicine, Zhejiang University, Hangzhou, Zhejiang, China

Wang Xi Interdisciplinary Institute of Neuroscience and Technology, The Second Affiliated Hospital, School of Medicine, Zhejiang University, Hangzhou, Zhejiang, China

Xiao-Hong Xu Institute of Neuroscience, State Key Laboratory of Neuroscience, CAS Center for Excellence in Brain Science and Intelligence Technology, Chinese Academy of Sciences, Shanghai, China

Shanghai Center for Brain Science and Brain-Inspired Intelligence Technology, Shanghai, China

Fen Yang Interdisciplinary Institute of Neuroscience and Technology, The Second Affiliated Hospital, School of Medicine, Zhejiang University, Hangzhou, Zhejiang, China

Yiwen Yang Department of Neurobiology, Key Laboratory for Biomedical Engineering of Education Ministry, Zhejiang University School of Medicine, HangzhouZhejiang, China

Department of Neurosurgery of Second Affiliated Hospital, Key Laboratory for Biomedical Engineering of Education Ministry, Zhejiang University School of Medicine, Hangzhou, Zhejiang, China

Zi-Xian Yu Institute of Neuroscience, State Key Laboratory of Neuroscience, CAS Center for Excellence in Brain Science and Intelligence Technology, Chinese Academy of Sciences, Shanghai, China

University of the Chinese Academy of Sciences, Beijing, China

Jiaozhen Zhang Department of Neurobiology, Key Laboratory for Biomedical Engineering of Education Ministry, Zhejiang University School of Medicine, Hangzhou, Zhejiang, China

Department of Neurosurgery of Second Affiliated Hospital, Key Laboratory for Biomedical Engineering of Education Ministry, Zhejiang University School of Medicine, Hangzhou, Zhejiang, China

Kuan Zhang Brain Research Center and State Key Laboratory of Trauma, Burns, and Combined Injury, Third Military Medical University, Chongqing, China

Wen Zhang National Institute on Drug Dependence, Peking University, Beijing, China

Yang Zhan Brain Cognition and Brain Disease Institute, Shenzhen Institutes of Advanced Technology, Chinese Academy of Sciences, Shenzhen, Guangdong, China

Neural Circuits Underlying Innate Fear

1

Chaoran Ren and Qian Tao

Abstract

Fear is defined as a fundamental emotion promptly arising in the context of threat and when danger is perceived. Fear can be innate or learned. Examples of innate fear include fears that are triggered by predators, pain, heights, rapidly approaching objects, and ancestral threats such as snakes and spiders. Animals and humans detect and respond more rapidly to threatening stimuli than to non-threatening stimuli in the natural world. The threatening stimuli for most animals are predators, and most predators are themselves prey to other animals. Predatory avoidance is of crucial importance for survival of animals. Although humans are rarely affected by predators, we are constantly challenged by social threats such as a fearful or angry facial expression. This chapter will summarize the current knowledge on brain circuits processing innate fear responses to visual stimuli derived from studies conducted in mice and humans.

C. Ren (✉)
Ministry of Education CNS Regeneration Collaborative Joint Laboratory, Guangdong-Hongkong-Macau Institute of CNS Regeneration, Jinan University, Guangzhou, China
e-mail: tchaoran@jnu.edu.cn

Q. Tao
School of Basic Medicine, Jinan University, Guangzhou, China

Keywords

Innate fear · Looming · Amygdala · Pulvinar

1.1 Introduction

Animals promote their survival by avoiding rapidly approaching objects that indicate threats. Looming stimulus-induced fear responses are conserved across species. For instance, expanding shadows specifying an impending collision can induce an avoidance response and upset in both infants and adults (Ball and Tronick 1971; King et al. 1992). In response to an expanding dark disk on a screen mimicking a predator, laboratory mice exhibit fear behaviors with either escape or freezing patterns (Yilmaz and Meister 2013). Given the robustness of looming stimuli-evoked fear behaviors, it is crucial to dissect the neural circuits that mediate this response.

1.2 Animal Studies

1.2.1 Retinal Ganglion Cells That Detect Looming Signals

Vision is the only useful sensory modality for initiating looming-evoked fear responses. It is well established that retinal ganglion cells (RGCs) are the final output neurons of the vertebrate retina, which collect visual information

H. Wang (ed.), *Neural Circuits of Innate Behaviors*, Advances in Experimental Medicine and Biology 1284, https://doi.org/10.1007/978-981-15-7086-5_1

from bipolar cells and amacrine cells. An organism as a whole cannot behaviorally respond to visual stimuli that are not also detectable by individual ganglion cells. Identifying the RGCs that can detect and transmit looming signals is a crucial step in understanding the neural basis of looming-evoked fear responses.

The light response patterns of RGCs are diverse. There are three types of signal detection in RGCs (Hartline 1938). ON-type signal detection results in a transient burst to light onset and a sustained elevated discharge rate throughout the photic stimulation. ON-OFF-type signal detection results in discharge bursts at both the onset and cessation of light stimuli. OFF-type signal detection is quiet until the stimulus light is turned off. There are two important components in a looming signal: dimming and motion; therefore, RGCs extracting this feature from the visual scene should be able to detect both stimuli. In accordance with these criteria, candidate RGC subtypes have been suggested in mice. For example, using genetic labeling, two-photon microscopy, and electrophysiology approaches, Münch et al. identified an approach-sensitive ganglion cell type in the mouse retina named PV-5 cells (Münch et al. 2009). The authors found that PV-5 cells belong to the OFF ganglion cell type, of which ~80% have dendrites that arborize in the inner plexiform layer (IPL). The spiking responses of PV-5 cells were evoked preferentially by stimuli mimicking approaching motion compared to either lateral motion or receding motion. Although the morphological and physiological features of PV-5 cells seem well positioned to detect looming signals, it remains to be determined whether PV-5 cells are necessary for looming-evoked fear responses in behaving animals. On the other hand, our recent study demonstrated that a looming stimulus can activate a previously undescribed subtype of RGC that innervates the dorsal raphe nuclei (DRNs) and superior colliculus (SC) (Huang et al. 2017). We found that dendrites of DRN/SC-projecting RGCs stratified in both the ON and OFF sublaminae of the IPL and that specific ablation of those RGCs through a saporin-based immunotoxin strategy impairs looming-evoked fear responses (freezing and escape behaviors), suggesting that those RGCs are necessary for looming-evoked fear responses. Although DRN/SC-projecting RGCs have an asymmetric dendritic field that resembles direction-selective RGCs, DRN/SC-projecting RGCs: (1) lack CART immunoreactivity and (2) show no direction preference to moving stimuli. It remains to be determined how looming stimuli activate DRN/SC-projecting RGCs that are nondirectional although directional summation in nondirection-selective RGCs has been described previously (Abbas et al. 2013).

1.2.2 Brain Circuits That Mediate Looming-Evoked Fear Responses in Mice

Looming signals detected by RGCs need to activate the brainstem fear systems to initiate fear responses. The precise circuits underlying such responses are not well understood. Accumulating evidence suggests that the SC, which is a retinal recipient structure, contributes to fear-related behaviors. For example, stimulation of SC neurons induces defensive behaviors (Sahibzada et al. 1986; Dean et al. 1988; Keay et al. 1988; Schenberg et al. 1990), and SC lesions impair defensive reactions to a sudden overhead visual stimulus (Dean et al. 1989). Therefore, if the SC receives looming-related signals transmitted from RGCs, it might be in a position to modulate looming-evoked fear responses. Consistent with this view, several circuits related to the SC have been proposed for mediating looming-induced fear behaviors. For instance, Wei et al. found that optogenetic activation of CaMKIIa neurons in the intermediate layer of the SC induced freezing-like behaviors, whereas silencing of those neurons reversibly blocked the expression of looming-evoked freezing (Wei et al. 2015). Furthermore, the authors demonstrated that looming-sensitive SC neurons can innervate the lateral posterior nucleus of the thalamus (LP), which in turn activates the basolateral amygdala (BLA) and that abrupt the signal transmission of this pathway impairs looming-evoked freezing.

Therefore, the authors provide compelling evidence that the SC-LP-BLA pathway plays a pivotal role in the regulation of looming-evoked freezing behaviors. In contrast, Shang et al. found that PV^+ excitatory neurons in the superficial layer of the SC can also detect looming signals and that specific activation of PV^+ SC neurons triggers escape-like behaviors (Shang et al. 2015). The authors also dissected the neural circuits underlying this process: PV^+ neurons in the SC can project to the parabigeminal nucleus (PBGN). Optogenetic activation of the PV^+ SC-PBGN pathway reliably induces escape behaviors. Furthermore, the authors prove that the PBGN can further innervate the central amygdala (CeA), which can also be activated by looming stimulation. Collectively, the work conducted by Shang et al. suggests that the SC-PBGN-CeA pathway underlies looming-evoked escape behaviors. Although the key neural circuits that initiate looming-evoked fear responses have been identified, another important question regarding the looming-evoked fear responses is how distinct defensive behaviors (i.e., freezing and escape) are selected by the brain. Shang et al. addressed this question by showing that SC orchestrates dimorphic fear behaviors with two divergent excitatory pathways (i.e., SC-LP and SC-PBGN) that work in a winner-take-all model (Shang et al. 2018). They proposed that general factors, including environmental context, threat stimulus features, and individual differences, determine behavioral patterns induced by looming stimuli.

Growing evidence suggests that changes in mood states can adjust looming-evoked fear responses, which is important for individual adaptations to challenges. Deciphering the neural circuits related to emotional centers that adjust looming-evoked fear responses will shed light on the mechanism of abnormal reactivity in mood disorders, such as anxiety, depression, and phobia. The monoaminergic systems derived from the midbrain DRN, locus coeruleus (LC), and ventral tegmental area (VTA) play a key role in the modulation of mood states. Changes in neural activity in the monoaminergic systems may influence the expression of fear responses through activation of related receptors distributed in fear regions, including the amygdala. In our previous study, we identified a retinoraphe projection with DRN-projecting RGCs that also send axonal collaterals to the SC (Huang et al. 2017). We demonstrated that looming signals can not only initiate fear responses by activating the retina-SC pathway but can also inhibit the serotonergic tone in the DRN, which can facilitate the induction of fear responses. Our finding suggests that the primary sensory input regulates itself via the DRN. The added synaptic delay in the circuit must clearly be outweighed by an adaptive advantage in such an important innate survival response. The role of the LC and VTA in the regulation of looming-evoked fear responses was investigated by Liping Wang's lab. Li et al. found that exposure to repeated stress caused anxiety-like behaviors accompanied by accelerated fear responses to looming stimulation (Li et al. 2018). The underlying neural mechanisms were investigated using an array of brain circuit interrogation tools, including c-Fos brain mapping, fiber photometry, chemogenetics, and optogenetics. They demonstrated that the LC-SC pathway is both necessary and sufficient for the stress-induced acceleration of looming-evoked fear responses. In addition, a very recent study conducted by Zhou et al. found that looming stimulation can also activate a subset of CaMKIIa+ neurons in the deep layer of SC, which could synapse onto CeA-projecting GABAergic neurons in the VTA (Zhou et al. 2019). Optogenetic activation of the SC-to-VTA projections induced escape behaviors, whereas inhibition of VTA GABAergic neurons impaired looming-evoked escape behaviors. These results demonstrated that visual circuits related to the VTA can also mediate looming-evoked fear responses. The precise interactions between the SC-VTA pathway identified by Zhou et al. and the SC-PBGN pathway identified by Shang et al. need to be determined (Zhou et al. 2019; Shang et al. 2015). One possible explanation of these redundant circuits for mediating looming-evoked escape behaviors is that the SC-PBGN pathway is dedicated to triggering active escape behaviors, whereas the SC-VTA pathway can also facilitate looming-evoked escape behaviors through modulation of the neural activity in the VTA.

Recent studies also found that mood-related brain regions other than the monoaminergic systems can also regulate looming-evoked fear responses. For example, Evans et al. found that a subset of excitatory neurons in the deep layer of the medial SC (mSC) can directly synapses onto the glutamatergic neurons in the dorsal periaqueductal gray (dPAG) (Evans et al. 2018). Changes in the activity of the mSC-dPAG pathway can regulate looming-evoked escape behavior. On the other hand, Salay et al. demonstrated that midline thalamic nuclei (i.e., the nuclei of the ventral midline thalamus, the xiphoid nucleus, and nucleus reuniens) can also regulate looming-evoked fear responses, including freezing, tail rattling, and autonomic arousal (Salay et al. 2018).

1.3 Human Studies

From the evolutionary point of view, innate fear subserves "self-protection" function that promotes the initiation of fight-or-flight response in the absence of awareness. It is therefore believed that initial responses triggered by innate fear are automatic and quick. This notion is supported by electrophysiological data demonstrating that detection of fear-related stimuli is as quickly as 100 ms post-stimulus or even earlier, which is more quickly than detection of non-fear stimuli (Mogg and Bradley 2010; Vuilleumier and Pourtois 2007).

The neural mechanism underlying innate fear is most likely involve non-conscious emotion processing. A host of techniques and experimental paradigms have been used to elicit non-conscious emotion processing. For instance, a backward masking procedure briefly presented an emotional stimulus (target) that is immediately followed by a masking emotional stimulus (mask), and it is most likely that the observer cannot consciously report the presence or the content of the target. Other techniques include binocular rivalry or flash suppression, during which the stimuli are presented at a subliminally threshold. The neuroimaging studies using such techniques have shown consistently that unseen stimuli of fear elicit activity in the amygdala. For instance, Whalen and colleagues presented pictures of fearful and happy facial expressions to healthy subjects by using the backward masking procedure and in meanwhile functional magnetic resonance imaging (fMRI) data was collected (Whalen et al. 1998). Although subjects reported seeing only neutral facial expressions, fMRI results found significant more activations in the amygdala during viewing of masked fearful faces than during viewing of masked happy faces. By recording intracranial electrophysiological data, a short-latency fear-related amygdala response was found during fearful, but not neutral or happy, facial expressions (Mendez-Bertolo et al. 2016). Another line of evidence comes from investigations on blindsight patients with striate cortex lesions, who could discriminate the content of emotional stimuli presented in their blind field. The results obtained in these patients also showed that the unseen fear stimuli increased amygdala activation, which were parallel with the data from healthy subjects when masking techniques were used. Some studies further demonstrate a correlation between their proficiency and activity in the amygdala (Pegna et al. 2005; Tamietto et al. 2009). These findings suggested that fear-related information can be perceived in the absence of awareness despite lesions to the visual cortical pathway.

Converging evidence suggests a subcortical pathway is underlying innate fear processing. First, it is evidenced that 5-month-old infants look longer at spiders than at non-threatening biological stimuli (e.g., flower), and 8-month-old infants responded more rapidly to snakes than to flowers and more rapidly to angry than to happy face (Rakison and Derringer 2008; LoBue and DeLoache 2010). Given that the infants have had little experience with the threatening stimuli, these results suggested that a subcortical mechanism underlying innate fear present from birth. Second, accumulated neuroimaging evidence suggests that the visual information of threat is transmitted from retina to amygdala via a subcortical pathway comprising the superior colliculus (SC) and pulvinar by demonstrating co-activation among these three brain regions in healthy subjects (Morris et al. 1999; Vuilleumier et al. 2003) as well as blindsight patients (Morris and Dolan 2001; Pegna et al. 2005) when they view fear-related stimuli. Dynamic causal modeling (DCM) is a powerful approach that is informed by anatomical and physiological principles to investigate effective connectivity between brain regions. Several DCM studies investigated whether the activation of these subcortical regions is causally related, and the studies have consistently showed a forward connection between the pulvinar and amygdala (McFadyen et al. 2017; Garvert et al. 2014; Rudrauf et al. 2008). Finally, lesion studies have shown that patients with unilateral pulvinar lesions impair discrimination of fearful faces in the contralateral fields (Ward et al. 2007). Furthermore, hemianopic patients without blindsight with pulvinar lesions demonstrated no facilitatory effects on detecting fearful faces, whereas hemianopic patients without pulvinar lesions showed response facilitation to fearful stimuli (Caterina et al. 2018). These findings suggest a pivotal role of pulvinar in implicit fear processing.

An important question is whether there is anatomical evidence that the SC-pulvinar-amygdala pathway exists in the human brain. Tamietto and colleagues used diffusion tensor imaging (DTI) to characterize in vivo the connectivity between the SC, pulvinar, and amygdala in ten healthy individuals and a blindsight patient with early unilateral destruction of the visual cortex (Tamietto et al. 2012). The authors found pulvinar-amygdala fiber connections and SC-pulvinar-amygdala fiber connections in the healthy individuals as well as the patient. Destruction of the visual cortex led to increased fiber connections along the subcortical pathway but only in the damaged hemisphere. This finding supports a functional role of the subcortical pathway in conveying visual emotional information critical for the blindsight patient. Similarly, Rafal et al. used probabilistic DTI tractography to reconstruct the subcortical pathway in both hemispheres for 19 of the 20 healthy human participants and 7 of the 8 macaques (Rafal et al. 2015). Importantly, it was evidenced that the microstructure of SC-amygdala pathway predicts threat bias, suggesting a functional role of the subcortical pathway in processing threat in healthy humans (Koller et al. 2018). The sample of human subjects was expanded in a multimodal neuroimaging study (McFadyen et al. 2019). The authors computationally modeled the hemodynamic activity during an emotion task and demonstrated a correlation between fiber density in this subcortical pathway and fearful face recognition as well as the strength of dynamic coupling between the SC, pulvinar, and amygdala.

In addition to the subcortical SC-pulvinar-amygdala pathway, some cortical regions may be also involved during non-conscious perception of fear. Neuroimaging studies in healthy humans suggest that fear-related stimuli selectively activate prefrontal, and orbitofrontal regions, anterior cingulate, and brain stem. The orbitofrontal cortex may extract threat value and inform threat perception via a feedback pathway to early visual regions (Barrett and Bar 2009; Kveraga et al. 2007).

In summary, an "innate alarm system" underlying perception of innate fear is a brain network comprising both subcortical and cortical regions. The amygdala seems to be a core site within this network. This network facilitates an immediate and fast response to threatening stimuli.

References

Abbas SY, Hamade KC, Yang EJ, Nawy S, Smith RG, Pettit DL (2013) Directional summation in non-direction selective retinal ganglion cells. PLoS Comput Biol 9:e1002969

Ball W, Tronick E (1971) Infant responses to impending collision: optical and real. Science 171:818–820

Barrett LF, Bar M (2009) See it with feeling: affective predictions during object perception. Philos Trans R Soc Lond Ser B Biol Sci 364:1325–1334

Caterina B, Mattia P, Davide B, Elisabetta L (2018) Pulvinar lesions disrupt fear-related implicit visual processing in Hemianopic patients. Front Psychol 9:2329

Dean P, Mitchell IJ, Redgrave P (1988) Responses resembling defensive behaviour produced by microinjection

of glutamate into superior colliculus of rats. Neuroscience 24:501–510
Dean P, Redgrave P, Westby GW (1989) Event or emergency? Two response systems in the mammalian superior colliculus. Trends Neurosci 12:137–147
Evans DA, Stempel AV, Vale R, Ruehle S, Lefler Y, Branco T (2018) A synaptic threshold mechanism for computing escape decisions. Nature 558:590–594
Garvert MM, Friston KJ, Dolan RJ, Garrido MI (2014) Subcortical amygdala pathways enable rapid face processing. NeuroImage 102(Pt 2):309–316
Hartline HK (1938) The response of single optic nerve fibers of the vertebrate eye to illumination of the retina. Am J Phys 121:400–415
Huang L, Yuan T, Tan M, Xi Y, Hu Y, Tao Q et al (2017) A retinoraphe projection regulates serotonergic activity and looming-evoked defensive behaviour. Nat Commun 8:14908
Keay KA, Redgrave P, Dean P (1988) Cardiovascular and respiratory changes elicited by stimulation of rat superior colliculus. Brain Res Bull 20:13–26
King SM, Dykeman C, Redgrave P, Dean P (1992) Use of a distracting task to obtain defensive head movements to looming visual stimuli by human adults in a laboratory setting. Perception 21:245–259
Koller K, Rafal RD, Platt A, Mitchell ND (2018) Orienting toward threat: contributions of a subcortical pathway transmitting retinal afferents to the amygdala via the superior colliculus and pulvinar. Neuropsychologia 128:78–86
Kveraga K, Boshyan J, Bar M (2007) Magnocellular projections as the trigger of top-down facilitation in recognition. J Neurosci 27:13232
Li L, Feng X, Zhou Z, Zhang H, Shi Q, Lei Z et al (2018) Stress accelerates defensive responses to looming in mice and involves a locus coeruleus-superior colliculus projection. Curr Biol 28:859–871
LoBue V, DeLoache JS (2010) Superior detection of threat-relevant stimuli in infancy. Dev Sci 13:221–228
McFadyen J, Mermillod M, Mattingley JB, Halász V, Garrido MI (2017) A rapid subcortical amygdala route for faces irrespective of spatial frequency and emotion. J Neurosci 37:3864–3874
McFadyen J, Mattingley JB, Garrido MI (2019) An afferent white matter pathway from the pulvinar to the amygdala facilitates fear recognition. elife 8:e40766
Mendez-Bertolo C, Moratti S, Toledano R, Lopez-Sosa F, Martinez-Alvarez R, Mah YH et al (2016) A fast pathway for fear in human amygdala. Nat Neurosci 19:1041–1049
Mogg K, Bradley B (2010) Orienting of attention to threatening facial expressions presented under conditions of restricted awareness. Cognit Emot 13:713–740
Morris J, Dolan R (2001) The amygdala and unconscious fear processing. In: De Gelder B, de Haan E, Heywood CA (eds) Out of mind: varieties of unconscious processes. Oxford Univ. Press, New York, pp 186–204
Morris JS, O¨ hman A, Dolan RJ (1999) A subcortical pathway to the right amygdala mediating bunseenQ fear. Proc Natl Acad Sci U S A 96:1680–1685
Münch TA, da Silveira RA, Siegert S, Viney TJ, Awatramani GB, Roska B (2009) Approach sensitivity in the retina processed by a multifunctional neural circuit. Nat Neurosci 12(10):1308–1316
Pegna AJ, Khateb A, Lazeyras F, Seghier ML (2005) Discriminating emotional faces without primary visual cortices involves the right amygdala. Nat Neurosci 8:24–25
Rafal RD, Koller K, Bultitude JH, Mullins P, Ward R, Mitchell AS et al (2015) Connectivity between the superior colliculus and the amygdala in humans and macaque monkeys: virtual dissection with probabilistic DTI tractography. J Neurophysiol 114:1947–1962
Rakison DH, Derringer J (2008) Do infants possess an evolved spider-detection mechanism? Cognition 107:381–393
Rudrauf D, David O, Lachaux JP, Kovach CK, Martinerie J, Renault B et al (2008) Rapid interactions between the ventral visual stream and emotion-related structures rely on a two-pathway architecture. J Neurosci 28:2793–2803
Sahibzada N, Dean P, Redgrave P (1986) Movements resembling orientation or avoidance elicited by electrical stimulation of the superior colliculus in rats. J Neurosci 6:723–733
Salay LD, Ishiko N, Huberman AD (2018) A midline thalamic circuit determines reactions to visual threat. Nature 557(7704):183–189
Schenberg LC, Costa MB, Borges PC, Castro MF (1990) Logistic analysis of the defense reaction induced by electrical stimulation of the rat mesencephalic tectum. Neurosci Biobehav Rev 14:473–479
Shang C, Liu Z, Chen Z, Shi Y, Wang Q, Liu S et al (2015) BRAIN CIRCUITS. A parvalbumin-positive excitatory visual pathway to trigger fear responses in mice. Science 348(6242):1472–1477
Shang C, Chen Z, Liu A, Li Y, Zhang J, Qu B et al (2018) Divergent midbrain circuits orchestrate escape and freezing responses to looming stimuli in mice. Nat Commun 9:1232
Tamietto M, Castelli L, Vighetti S, Perozzo P, Geminiani G, Weiskrantz L et al (2009) Unseen facial and bodily expressions trigger fast emotional reactions. Proc Natl Acad Sci USA 106:17661–17666
Tamietto M, Pullens P, de Gelder B, Weiskrantz L, Goebel R (2012) Subcortical connections to human amygdala and changes following destruction of the visual cortex. Curr Biol 22:1449–1455
Vuilleumier P, Pourtois G (2007) Distributed and interactive brain mechanisms during emotion face perception: evidence from functional neuroimaging. Neuropsychologia 45:174–194
Vuilleumier P, Armony JL, Driver J, Dolan RJ (2003) Distinct spatial frequency sensitivities for processing faces and emotional expressions. Nat Neurosci 6:624–631

Ward R, Calder AJ, Parker M, Arend I (2007) Emotion recognition following human pulvinar damage. Neuropsychologia 45:1973–1978
Wei P, Liu N, Zhang Z, Liu X, Tang Y, He X et al (2015) Processing of visually evoked innate fear by a non-canonical thalamic pathway. Nat Commun 6:6756
Whalen PJ, Rauch SL, Etcoff NL, McInerney SC, Lee MB, Jenike MA (1998) Masked presentations of emotional facial expressions modulate amygdala activity without explicit knowledge. J Neurosci 18:411–418
Yilmaz M, Meister M (2013) Rapid innate defensive responses of mice to looming visual stimuli. Curr Biol 23:2011–2015
Zhou Z, Liu X, Chen S, Zhang Z, Liu Y, Montardy Q et al (2019) A VTA GABAergic neural circuit mediates visually evoked innate defensive responses. Neuron 103:473–488.e6

Neurobiology and Neural Circuits of Aggression

2

Mohamed Helmy, Jiaozhen Zhang, and Hao Wang

Abstract

Aggression takes several forms and can be offensive or defensive. Aggression between animals of the same species or society aims to inflict harm upon another for the purpose of protecting a resource such as food, reproductive partners, territory, or status. This chapter explores the neurobiology of aggression. We summarize the behavior of aggression, rodent models of aggression, and the correlates of aggressive behavior in the context of neuroendocrinology, neurotransmitter systems, and neurocircuitry. Translational implications of rodent studies are briefly discussed, applying basic research to brain imaging data and therapeutic approaches to conditions where aggression is problematic.

Keywords

Aggression · Neural circuits · Hypothalamus

2.1 Introduction

Aggression can be defined as delivering harm to another, with the aim of taking advantage of a limited resource (Haller 2018a). Motivation for aggression in animals can be: (i) offensive, which occurs during competition for, or protection of, resources such as food, reproductive partners, social status, or territory; (ii) defensive, to ward off attacks from another animal of the same or different species; (iii) maternal, to protect offspring; (iv) predatory, or hunting, capturing, and consuming prey; (v) play-fighting, shown by adolescent individuals; and (vi) patrol or marking as a form of agnostic behavior (Adams 2006; Veenema 2009). Differences in animal societal levels of aggression can be attributed to increased mating aggression and decreased parental investment (Barber 2008).

Offensive or predatory aggression is expressed in humans as instrumental or proactive aggression, where some goal is aimed for, whereas defensive aggression corresponds to impulsive or reactive aggression in humans, and occurs in response to perceived attack or threat (Blair

M. Helmy · J. Zhang
Department of Neurobiology and Department of Neurosurgery of Second Affiliated Hospital, Key Laboratory for Biomedical Engineering of Education Ministry, Zhejiang University School of Medicine, Hangzhou, Zhejiang, China

H. Wang (✉)
Department of Neurobiology and Department of Neurosurgery of Second Affiliated Hospital, Key Laboratory for Biomedical Engineering of Education Ministry, Zhejiang University School of Medicine, Hangzhou, Zhejiang, China

NHC and CAMS Key Laboratory of Medical Neurobiology, MOE Frontier Science Center for Brain Research and Brain Machine Integration, School of Brain Science and Brain Medicine, Zhejiang University, Hangzhou, Zhejiang, China
e-mail: haowang@zju.edu.cn

H. Wang (ed.), *Neural Circuits of Innate Behaviors*, Advances in Experimental Medicine and Biology 1284, https://doi.org/10.1007/978-981-15-7086-5_2

2016). Aggression is also associated with lack of empathy (Hernandez-Lallement et al. 2018).

This chapter reviews the neurobiology of aggression, recent findings in aggression research in rodents, and summarizes its translational implications.

Fighting between conspecifics usually follows rules, such as signaling intent to allow for a weaker opponent to withdraw, lunge-and-bite attacks on relatively robust body parts such as rump, while avoiding non-vital body parts such as face and neck (this is not always the case), and cessation of violence when signals of defeat are expressed by the opponent (Adams 2006; Haller 2017). Abnormal or maladaptive aggression is quantitative or qualitative increase in normal aggressive behavior compared to controls, or departure from these species-specific expressions of violence or its intent (Haller and Kruk 2006; Miczek et al. 2015).

Some animals may seek aggression in operant and place preference paradigms, and this aggression seeking shows features common with compulsive and addictive behavior models, such as resistance to abstinence conditioning (Golden et al. 2017a, b). Previous experience in winning or losing fights partially determines outcome of future escalations to contact violence, in addition to multiple other factors, such as prior residency (Hsu et al. 2006). Male animals with a history of winning fights show increased aggression after a period of deprivation from fighting opportunities (Kudryavtseva et al. 2011).

Aggression in animals may show seasonal variations during reproductive periods and to protect territory, and this is associated with changes in circulating gonadal steroids (Munley et al. 2018). Aggression toward prey, or hunting, is obviously behaviorally distinct from that directed at conspecifics.

2.2 Experimental Paradigms to Explore Aggression in Rodents

Resident-intruder paradigm: a male rodent, the resident, is allowed to familiarize with a home cage, with or without a female. An intruder animal, also male, is then introduced to the resident. A fight ensues, in which, all other conditions being relatively equal, the resident is expected to defeat the intruder.

Maternal aggression: a lactating dam with pups actively defends her nest and pups against an intruder, an unfamiliar male or female animal. Attacks can be fierce, may target face and neck of the intruder, and are quickly initiated.

Predatory aggression: a rodent is allowed to attack and consume prey, usually an insect.

Aggression seeking: This is measured using conditioned place preference after exposure of an animal to a conspecific in one compartment for some days, or in an operant setup where a lever is pressed or a nose poke is required to gain access to a conspecific on which to aggress.

Aggression inheritance in rodents is polygenic with a wide variety of strains in which to model, in addition to considering developmental factors such as maternal care, and the experimental paradigm used to test aggression; animal selection for modeling aggression therefore requires scrutiny (Miczek et al. 2001; Natarajan et al. 2009; Nyberg et al. 2004).

Abnormal or excessive aggression can be modeled in rodents through: (i) stress models, which include repeated prolonged maternal separation and early isolation or social subjugation; (ii) drug models such as administration of anabolic steroids during adolescence or alcohol in adulthood; (iii) genetic, by selective breeding of animals showing high aggression or anxiety; and (iv) decreasing circulating glucocorticoid by adrenalectomy and low-dose corticosterone pellet implantation (Haller 2017; Takahashi et al. 2012). Early social deprivation is associated with increased aggression and attack behavior on vulnerable body parts (Tóth et al. 2008).

Exposure to chronic ultrasonic noise increased aggression in the resident-intruder paradigm only in animals showing initial high levels of aggression in one series of experiments, but in all animals in another (Gorlova et al. 2019; Pavlov et al. 2017). Surgical devocalization of rats increased propensity to aggressive behavior during neutral interactions (Kisko et al. 2017).

2.3 Neuroendocrinology of Aggression

During development: Maternal separation is associated with increased play-fighting in adolescence and aggression in adulthood, and increased basal corticosterone and hypothalamic vasopressin (Veenema and Neumann 2009). Impacting dam–pup interaction, including reduced bedding material, is associated with higher circulating corticosteroids, lower corticotropic hormone, and increased aggression (Rice et al. 2008). Adolescent animals attacked by an adult show changes in the vasopressinergic system, namely vasopressinergic fibers are increased and serotoninergic terminals are decreased (Ferris 2000). Down-regulated or impaired oxytocinergic activity is associated with increased aggression, but exogenous administration does not ameliorate aggressive behavior (de Jong and Neumann 2018).

In Males: Testosterone as a biological root of aggression is contested (Albert et al. 1993). Deletion of androgen receptors in the nervous system is associated with impaired display of masculine behavior (Juntti et al. 2010). Development of play-fighting is dependent on androgen- and estrogen-mediated effects (Field et al. 2006). Higher aggression in male animals was correlated with higher adrenocorticotropin hormone responsiveness, lower trait anxiety, and great Fos immunoreactivity in paraventricular nucleus (Veenema et al. 2007).

In Females: Maternal aggression (actions mediated by a pregnant or lactating dam against others) is modulated by ovarian steroids, stimulated by suckling pups, and increased if the pups are handled (de Almeida et al. 2014; Giovenardi et al. 2005). During estrus in a lactating dam, a male intruder is either attacked or solicited, and this may be associated with changes in perception of male-specific urinary proteins (Agrati et al. 2011; Martín-Sánchez et al. 2015). However, male-specific urinary proteins alone do not instigate aggression in animals (Mucignat-Caretta et al. 2004; however see also Chamero et al. 2007), indicating that other sensory stimuli are also involved. Vasopressin release in central amygdala and oxytocin release in central amygdala and paraventricular nucleus contribute to maternal aggression (Bosch and Neumann 2010; Bosch et al. 2005; Bosch 2013). Female aggression has not received the same attention as male aggression; due to ethological background, neurocircuitry of femal aggression is better investigated in rodent models other than C57BL/6 mice, such as Swiss Webster mice, rats, and Syrian hamsters (Been et al. 2019).

2.4 Neurotransmitter Systems in Aggression

Neurotransmitters directly implicated in physiology and pathology of aggression include serotonin, dopamine, and GABA (de Almeida et al. 2005). Factors altering serotonin and dopamine neurotransmission during development are associated with life-long behavioral alteration (de Almeida et al. 2005).

Low brain serotonin has been correlated with high aggression but mechanistic explanations or a direct relationship is not agreed upon. Serotonin levels were lower in some brain areas of animals showing high aggression, and these same individual animals show the greatest increase in serotonin after the stress of agonistic behavior (Summers et al. 2005). Treatment with systemic specific serotonin agonists and antagonists suggests that normal and abnormal aggressive behaviors are mediated by different serotonin subtypes and mediate different types, phasic vs. tonic, of serotoninergic activity (de Boer and Koolhaas 2005). Chronically enhanced activity of 5-HT_{1A} serotonin autoreceptors is associated with increased aggression (Caramaschi et al. 2007).

It was suggested that aggression has different phases including appetitive and executive phases, which were influenced by pre- and post-synaptic serotoninergic neurotransmission respectively (Olivier and van Oorschot 2005). Modulation of aggression by serotoninergic neurotransmission interacts with glucocorticoids released during stress in a phase-, context-, and history-dependant manner (Summers and Winberg 2006). Studies have shown that reduced serotoninergic activity in prefrontal cortex is associated with altered serotonin receptor expression in forebrain, poorly regulated dopamine secretion in nucleus accumbens and aggressive impulsivity (Nautiyal et al. 2015; Niederkofler et al. 2016; Seo et al. 2008). Knockout of serotonin autoreceptors 1B decreased serotonin in brain and spinal cord regions probably through increased serotonin turnover, increased dopamine turnover in nucleus accumbens, decreased dopamine, and was associated with increased aggressiveness and heightened cocaine sensitivity (Ase et al. 2008). Increased alcohol-induced aggression is associated with expression of certain $GABA_A$ receptor subtypes, whereas activation of serotonin receptor subtypes is associated with decreased alcohol-induced aggression (Miczek et al. 2006).

Dopamine in nucleus accumbens increased in anticipation of aggression and peaked after confrontation, while serotonin in prefrontal cortex decreased in association with termination (Ferrari et al. 2003; van Erp and Miczek 2000). Lack of monoamine oxidase A but not B is associated with increased aggression (Cases et al. 1995; Shih et al. 1999).

Lack of endothelial nitric oxide synthase greatly decreases aggression in male mice but does not influence maternal aggression, whereas deficiency of neuronal nitric oxide synthase decreases maternal aggression in female mice (Demas et al. 1999; Gammie and Nelson 1999; Gammie et al. 2000). Lack of neuronal nitric oxide synthase increased aggression, decreased social investigation, and was associated with decreased serotonin turnover and deficient serotoninergic receptors (Chiavegatto et al. 2001; Trainor et al. 2007a).

2.5 Neurocircuitry of Aggressive Behavior

2.5.1 Amygdala

The amygdala plays a vital role in mediating many aspects of innate and learned emotional behaviors such as fear-conditioning, predation, and aggression. In rat muricide model, c-Fos immunoreactivity increased in medial, central, and basolateral amygdala, as well as lateral hypothalamus; periaqueductal gray activations shifted from dorsal to ventral columns (Tulogdi et al. 2015).

Estrogen receptors α and β in medial preoptic area and medial amygdala differentially modulate aggressive behavior in males (Nakata et al. 2016). Site-specific knockdown of an estrogen receptor β gene in medial preoptic area decreased aggressive but not sexual behavior in adulthood; in amygdala knockdown of estrogen receptor β and α did not impact aggression (*ibid*). GABAergic neurons in medial amygdala promote aggression and are inhibited by neighboring glutamatergic neurons; the latter promote solitary grooming (Hong et al. 2014).

Aromatase expressing neurons in posterodorsal medial amygdala modulate intermale aggression and maternal aggression (Unger et al. 2015). Estrogen-dependent gene expression increased in bed nucleus of stria terminalis during long days and was associated with decreased aggression (Laredo et al. 2014; Trainor et al. 2007b).

Protein expression in oxytocin- and vasopressin-positive neurons in hypothalamus and bed nucleus of stria terminalis after intermale aggression is associated with medial amygdalar connectivity and activity (Wang et al. 2013).

Increased early growth response factor 1 in medial amygdala was associated with increased maternal aggression (Hasen and Gammie 2006).

2.5.2 Hypothalamus

For decades, it was generally accepted that the aggression center in the brain is the hypothalamic attack area, which is located in mediobasal hypothalamus and receives inputs from medial prefrontal neurons, septal regions, bed nucleus of stria terminalis, medial amygdala, amygdalo-hippocampal subiculum, locally from hypothalamus, and from lateral parabrachial nucleus (Toth et al. 2010). Electrical stimulation of the hypothalamic attack area promptly induced attack in cats and rodents, and increased c-Fos immunoreactivity in the lateral septum, bed nucleus of stria terminalis, medial and central amygdala, mediodorsal thalamic nucleus, and piriform and cingular cortex (Halász et al. 2002), suggesting these brain areas are also involved in aggression. Fos immunoreactivity in medial preoptic area and nucleus accumbens correlates with mating and experience of aggression (McHenry et al. 2016). Bilateral lesions of medial preoptic area are associated with attenuated aggression (Albert et al. 1986). Maternal aggression in lactating mice toward a male intruder is associated with increased Fos immunoreactivity in medial preoptic area, extended amygdala, accessory olfactory bulb, claustrum, and other brain regions (Gammie and Nelson 2001; Hasen and Gammie 2005).

In 2011, the ventrolateral part of ventromedial hypothalamus (VMHvl) was identified as a node structure to initiate attack (Lin et al. 2011). Optogenetic activation of neurons in VMHvl initiated male attack against conspecifics and females, as well as inanimate objects. Single unit activity measured widespread activation during aggressive encounters, but low and diminishing activity during mating (*ibid*). Consistently, single unit recording showed that the activity of VMHvl neurons is correlated with investigating olfactory cues of male conspecifics and attack; neuronal activity increased as the male–male distance decreased (Falkner et al. 2014). Optogenetic activation of VMHvl potentiated aggression-seeking and attack ferocity; inhibition had an opposite effect (Falkner et al. 2016).

Within the VMHvl, a group of estrogen receptor α-positive neurons has been shown to play an essential role in aggression. Calcium activity and optogenetic activation of estrogen receptor α-positive (Esr1+) neurons in the anterior part of VMHvl were associated with defense behavior against a conspecific, including non-threatening female; optogenetic inhibition impaired defense behavior against an aggressive conspecific (Wang et al. 2019a). Fos immunoreactivity shows overlapping hypothalamic and amygdalar activation after an aggressive or sexual encounter in males, indicating aggression and mating behaviors may be regulated by the same type of neurons (Veening et al. 2005). This concept is supported by a recent study showing that increasing photostimulation power on VMHvl Esr1+ neurons in male mice shifted behavior during a single interaction with a male or female mouse from investigation, to mounting, to attack, whereas non-cell-specific optogenetic activation in the same area was associated with attack but not mounting (Lee et al. 2014). Collectively, these results suggested that the VMHvl Esr1+ neurons coordinate scalable control of two distinct behaviors, namely aggression and mating.

Interestingly, roles of VMHvl Esr1+ neurons seem to be varied in different mouse strains and sexual history. Optogenetic activation of VMHvl Esr1+ neurons in virgin C57 female mice was associated with attack on an intruder female mouse, whereas activation of the same cells in virgin Swiss Webster and lactating C57 female mice was associated with mounting of a female intruder (Hashikawa et al. 2017). Unexpectedly, knockdown of estrogen receptor α in ventromedial hypothalamus increased female aggression against juveniles (Spiteri et al. 2010). In addition to Esr1+ neurons, other cell types in VMH are also involved in aggression. For example, genetic ablation of progesterone expressing neurons in ventromedial hypothalamus inhibited sexual receptivity in females and mating and aggression in males (Yang et al. 2013). In addition, a subset of neurons in VMHvl has been shown to mediate social fear (Sakurai et al. 2016).

Pheromone and olfactory receptors participate in hypothalamic circuits modulating aggressive behavior, but comprehensive descriptions are lacking (Sternson 2013). Deficiency of TRP2-expressing neurons in vomeronasal organ impairs intermale aggression and sex discrimination (Stowers 2002). A component of urine acts as a phermone to instigate intermale aggression in mice *via* vomeronasal organ neuronal circuits (Chamero et al. 2007).

2.5.3 Prefrontal Cortex

Post-weaning social isolation was associated with increased aggression in later life, reduced prefrontal cortical thickness, and was associated with abnormal aggressive behavior such as reduced signaling and attack of vulnerable body areas (Biro et al. 2017). Optogenetic activation of excitatory neurons decreased the intensity of an aggressive bout and the propensity to initiate it, but was not associated with change in aggression termination; optogenetic suppression was associated with opposite effects (Takahashi et al. 2014).

2.5.4 Lateral Septum

Projections from lateral septum to ventrolateral part of ventromedial hypothalamus are inhibitory, and photoactivation of lateral septum cells terminates attack behavior (Wong et al. 2016). Loss of a calcium-activated chloride channel in a subpopulation of lateral septum neurons increased aggressive display in the resident-intruder paradigm (Wang et al. 2019b).

$GABA_A$ receptor agonist injected into lateral septum increased aggression (McDonald et al. 2012).

Animals bred for short attack latency or high anxiety behavior are more aggressive, and show reduced vassopressinergic neurotransmission in lateral septum, in addition to altered hypothalamo-pituitary-adrenal axis response and increased serotoninergic neurotransmission (Veenema and Neumann 2007). Vasopressin released in lateral septum modulates social behavior but not aggression (Beiderbeck et al. 2007).

2.5.5 Other Brain Areas

Winning fights increased neurogenesis in hippocampus and aggression in males, and decreased Fos immunoreactivity in amygdala; these effects discontinue with absence of further opportunities to aggress (Smagin et al. 2015). Dopamine decreased in nucleus accumbens in anticipation of an aggressive episode (Ferrari et al. 2003). Knockdown of progestin receptors in ventral tegmental area in female mice is associated with increased male rejection and aggression (Frye et al. 2014). Neural activity in dorsal midbrain central gray is associated with offensive and defensive aggressive behavior (Adams 2006). pCREB-positive cells increased in caudal periaqueductal gray and lateral septum after maternal aggression (Gammie and Nelson 2001).

2.5.6 Synthesis

Research summarized above suggests environmental signals to aggress are relayed through one of two possible main systems as shown in Fig. 2.1: (i) medial amygdala to extended amygdala, lateral septum, and hypothalamic areas, and therefrom to periaqueductal gray; (ii) hypothalamic attack area, the ventrolateral part of the ventromedial hypothalamus coordinating afferent and efferent brain signaling in the initiation and processing of aggression with afferents and efferents from and to prefrontal cortex, lateral septum, amygdala, other areas of hypothalamus, and brainstem (Aleyasin et al. 2018; de Boer et al. 2015; Nelson and Trainor 2007). Olfactory cues obviously play an important role in rodent social dynamics, but their contribution to human aggression is unknown. Signals from prefrontal cortex modulate or inhibit aggressive behavior. Data from human studies support a central role for amygdala, hypothalamus, and periaqueductal gray for processing and initiating aggressive impulses, and prefrontal cortex in processing action values and the decision to aggress (Blair 2016).

2.6 Translational Implications

Aggression in human history is obviously complex, and attempts at interpretation are placed in a socioeconomic context (Fortman and Bas de 2005).

Human aggression can be reactive or impulsive associated with anger and autonomic arousal, and instrumental which is thought to be more goal-oriented and involves less autonomic arousal; similarly aggression can be clustered into impulsive-affective and controlled-predatory subtypes (Nelson and Trainor 2007; Vitiello and Stoff 1997). Human conditions in which aggressive behavior is problematic include antisocial

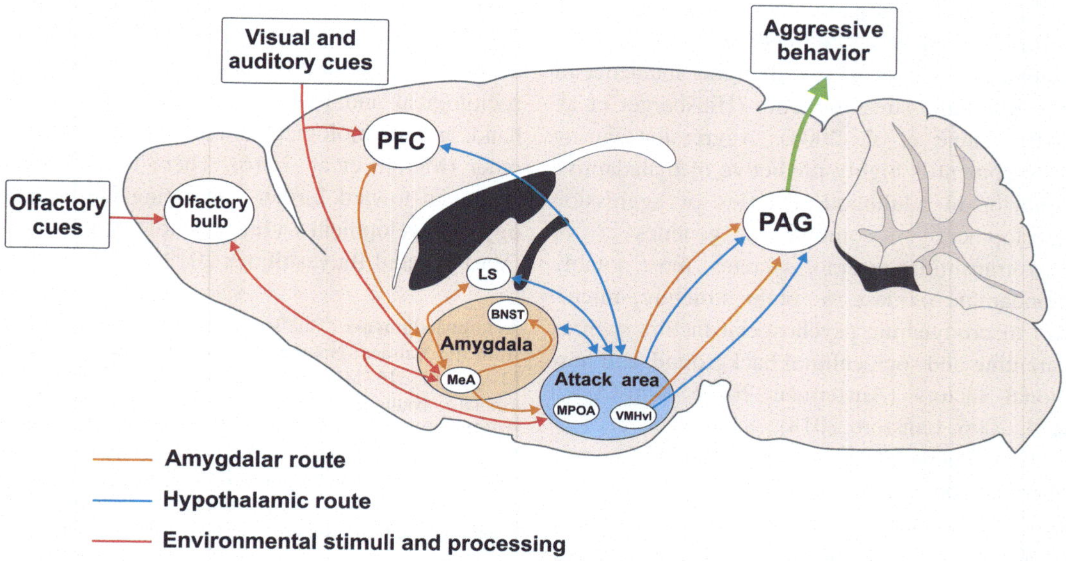

Fig. 2.1 Models of brain regions and circuitry mediating aggression. Neurocircuitry mediating aggressive behavior is processed either mainly through MeA, to BNST, LS, and hypothalamic areas, or through the hypothalamic attack area and VMHvl which communicates with amygdala, PFC, and LS. The output of these pathways is the PAG. Environmental stimuli of aggression are mostly social in nature. Olfactory cues are received by the olfactory bulb and relayed to MeA, hypothalamic areas, or both. Visual and auditory cues are processed by MeA as well as PFC. *BNST* bed nucleus of the stria terminalis, *LS* lateral septum, *MeA* medial amygdala, *MPOA* medial preoptic area, *PAG* periaqueductal gray, *PFC* prefrontal cortex

personality disorder, borderline personality disorder, intermittent explosive disorder, post-traumatic stress disorder, irritable and depression-linked aggression, schizophrenia, bipolar disorder, attention-deficit hyperactivity disorder, dementia and associated illness, and alcohol-related aggression (Coccaro et al. 2011; Nelson and Trainor 2007; Pompili et al. 2017). It was suggested that aggressive conditions can be classified into those associated with neurocircuitry pathology, hypoarousal and low circulating glucocorticoids, and emotional and physiological hyperarousal (Haller and Kruk 2006). Establishing animal model validity as related to human conditions is suggested (Haller 2018b).

Studies in animals and humans suggest neurocircuitry underlying aggression involves subcortical systems producing aggressive impulses, circuits predicting outcome of aggressing and making a decision to aggress or not, and notably circuits in prefrontal and medial temporal cortex regulating emotion (Anderson et al. 1999; Bufkin and Luttrell 2005; Coccaro et al. 2011; Davidson 2000). Aggression in humans associated with neurological damage or degeneration includes frontotemporal lesions, epilepsy, and Alzheimer's disease (Haller and Kruk 2006). Violent behavior in humans is associated with functional impairments in prefrontal cortex (Yang and Raine 2009). Bilateral amygdalar destruction in humans to treat intractable aggression has a "taming effect" but does not abolish aggressive outbursts (Lee et al. 1998). Deep brain stimulation of posterior medial hypothalamus and nucleus accumbens reduced aggression (Harat et al. 2015).

Borderline personality disorder is associated with volume loss in amygdala, hippocampus, and left orbitofrontal and right anterior cingulate cortex (van Elst et al. 2003). Psychopathy is associated with hypoactive frontolimbic circuit and hippocampal asymmetry (Raine et al. 2004; Veit et al. 2002).

Early to exposure to violence and cruelty may consolidate an attraction to aggression in an individual, and is associated with higher incidence of post-traumatic stress disorder (Hinsberger et al. 2016; Raine et al. 2004). Aggression during development is highly predictive of maladaptive behavior in adulthood; theories of aggression development combine genetics of neurotransmitter-receptor systems, most notably monoamine oxidase A, brain structure, micro- and macrodynamic psychosocial factors such as parenting and sociocultural background, and hormonal factors (Austerman 2017; Kim-Cohen et al. 2006; Lansford 2018).

Experiments in rodents are needed to improve pharmacotherapy of aggressive conditions, and which presently includes the following: (i) atypical antipsychotics such as clozapine and risperidone; (ii) anticonvulsants such as topiramate; (iii) mood stabilizers such as lithium; (iv) adrenergic receptor agonists such as clonidine; (v) typical antipsychotics such as haloperidol; (vi) benzodiazepines such as midazolam; (vii) combinations of drugs which may include histamine blockers such as promethazine; (viii) drugs acting on brain serotonin notably selective serotonin reuptake inhibitors; (ix) beta blockers such as propranolol; (x) drugs acting on nicotine receptors (Brieden et al. 2002; Buitelaar et al. 2001; Granic 2014; Hoptman 2015; Huf et al. 2016; Knapp et al. 2012; Pompili et al. 2017; Robb et al. 2019; Swann 2003). Studies in animals and humans show that oxidative stress plays an important role in alcohol toxicity and aggressive behavior (Tobore 2019).

In short, we have a wealth of data spanning across genetics, neurophysiology, brain structure, pharmacology, and behavior from animal and human studies on aggression; a comprehensive integration of this data may provide novel insights into how we can better screen for and manage conditions associated with or predisposing to aggression early in life, and shed light on common and disparate mechanisms underlying this complex phenomenon. Further integrating neuroscientific findings into the broader context of society would need a meaningful science of neurosociology, which has not yet developed a common language between the sociological and phenomenological on the one hand, and the biological and deterministic on the other (Meloni et al. 2016). There is reason for optimism toward further integrating the neurology of development within a sociological context (Vasileva and Balyasnikova 2019).

Acknowledgments This work was supported by grants from the National Natural Science Foundation of China (31970940, 31671100, and 31622027), the Zhejiang Provincial Natural Science Foundation of China (LR18H090001), the Non-profit Central Research Institute Fund of the Chinese Academy of Medical Sciences (2018PT31041), the Program for Introducing Talents in Discipline to Universities, and the Fundamental Research Funds for the Central Universities (2019QNA5001).

References

Adams DB (2006) Brain mechanisms of aggressive behavior: an updated review. Neurosci Biobehav Rev 30 (3):304–318. https://doi.org/10.1016/j.neubiorev.2005.09.004

Agrati D, Fernández-Guasti A, Ferreño M, Ferreira A (2011) Coexpression of sexual behavior and maternal aggression: the ambivalence of sexually active mother rats toward male intruders. Behav Neurosci 125 (3):446–451. https://doi.org/10.1037/a0023085

Albert DJ, Walsh ML, Gorzalka BB, Mendelson S, Zalys C (1986) Intermale social aggression: suppression by medial preoptic area lesions. Physiol Behav 38 (2):169–173. https://doi.org/10.1016/0031-9384(86)90151-4

Albert DJ, Walsh ML, Jonik RH (1993) Aggression in humans: what is its biological foundation? Neurosci Biobehav Rev 17(4):405–425

Aleyasin H, Flanigan ME, Russo SJ (2018) Neurocircuitry of aggression and aggression seeking behavior: nose poking into brain circuitry controlling aggression. Curr Opin Neurobiol Neurobiol Behav 49:184–191. https://doi.org/10.1016/j.conb.2018.02.013

de Almeida RM, Ferrari PF, Parmigiani S, Miczek KA (2005) Escalated aggressive behavior: dopamine, serotonin and GABA. Eur J Pharmacol 526(1–3):51–64. https://doi.org/10.1016/j.ejphar.2005.10.004

de Almeida RM, Ferreira A, Agrati D (2014) Sensory, hormonal, and neural basis of maternal aggression in rodents. In: Miczek KA, Meyer-Lindenberg A (eds) Neuroscience of aggression, Current Topics in Behavioral Neurosciences. Springer Berlin Heidelberg,

Berlin, Heidelberg, pp 111–130. https://doi.org/10.1007/7854_2014_312

Anderson SW, Bechara A, Damasio H, Tranel D, Damasio AR (1999) Impairment of social and moral behavior related to early damage in human prefrontal cortex. Nat Neurosci 2(11):1032–1037. https://doi.org/10.1038/14833

Ase AR, Reader TA, Hen R, Riad M, Descarries L (2008) Altered serotonin and dopamine metabolism in the CNS of serotonin 5-HT(1A) or 5-HT(1B) receptor knockout mice. J Neurochem 75(6):2415–2426. https://doi.org/10.1046/j.1471-4159.2000.0752415.x

Austerman J (2017) Violence and aggressive behavior. Pediatr Rev 38(2):69–80. https://doi.org/10.1542/pir.2016-0062

Barber N (2008) Evolutionary social science: a new approach to violent crime. Aggress Violent Behav 13 (3):237–250. http://www.sciencedirect.com/science/article/pii/S1359178908000190

Been LE, Gibbons AB, Meisel RL (2019) Towards a neurobiology of female aggression. Neuropharmacology 156:107451. https://doi.org/10.1016/j.neuropharm.2018.11.039

Beiderbeck DI, Neumann ID, Veenema AH (2007) Differences in intermale aggression are accompanied by opposite vasopressin release patterns within the septum in rats bred for low and high anxiety. Eur J Neurosci 26(12):3597–3605. https://doi.org/10.1111/j.1460-9568.2007.05974.x

Biro L, Toth M, Sipos E, Bruzsik B, Tulogdi A, Bendahan S, Sandi C, Haller J (2017) Structural and functional alterations in the prefrontal cortex after post-weaning social isolation: relationship with species-typical and deviant aggression. Brain Struct Funct 222(4):1861–1875. https://doi.org/10.1007/s00429-016-1312-z

Blair RJR (2016) The neurobiology of impulsive aggression. J Child Adolesc Psychopharmacol 26(1):4–9. https://doi.org/10.1089/cap.2015.0088

de Boer SF, Koolhaas JM (2005) 5-HT1A and 5-HT1B receptor agonists and aggression: a pharmacological challenge of the serotonin deficiency hypothesis. Eur J Pharmacol 526(1–3):125–139. https://doi.org/10.1016/j.ejphar.2005.09.065

de Boer SF, Olivier B, Veening J, Koolhaas JM (2015) The neurobiology of offensive aggression: revealing a modular view. Physiol Behav 146(July):111–127. https://doi.org/10.1016/j.physbeh.2015.04.040

Bosch OJ (2013) Maternal aggression in rodents: brain oxytocin and vasopressin mediate pup defence. Philos Trans R Soc Lond B Biol Sci 368(1631):20130085. https://doi.org/10.1098/rstb.2013.0085

Bosch OJ, Neumann ID (2010) Vasopressin released within the central amygdala promotes maternal aggression. Eur J Neurosci 31(5):883–891. https://doi.org/10.1111/j.1460-9568.2010.07115.x

Bosch OJ, Meddle SL, Beiderbeck DI, Douglas AJ, Neumann ID (2005) Brain oxytocin correlates with maternal aggression: link to anxiety. J Neurosci Off J Soc Neurosci 25(29):6807–6815. https://doi.org/10.1523/JNEUROSCI.1342-05.2005

Brieden T, Ujeyl M, Naber D (2002) Psychopharmacological treatment of aggression in schizophrenic patients. Pharmacopsychiatry 35(3):83–89. https://doi.org/10.1055/s-2002-31523

Bufkin JL, Luttrell VR (2005) Neuroimaging studies of aggressive and violent behavior: current findings and implications for criminology and criminal justice. Trauma Violence Abuse 6(2):176–191. https://doi.org/10.1177/1524838005275089

Buitelaar JK, van der Gaag RJ, Cohen-Kettenis P, Melman CT (2001) A randomized controlled trial of risperidone in the treatment of aggression in hospitalized adolescents with subaverage cognitive abilities. J Clin Psychiatry 62(4):239–248. https://doi.org/10.4088/jcp.v62n0405

Caramaschi D, de Boer SF, Koolhaas JM (2007) Differential role of the 5-HT1A receptor in aggressive and non-aggressive mice: an across-strain comparison. Physiol Behav 90(4):590–601. https://doi.org/10.1016/j.physbeh.2006.11.010

Cases O, Seif I, Grimsby J, Gaspar P, Chen K, Pournin S, Müller U, Aguet M, Babinet C, Shih JC (1995) Aggressive behavior and altered amounts of brain serotonin and norepinephrine in mice lacking MAOA. Science (New York, N.Y.) 268(5218):1763–1766. https://doi.org/10.1126/science.7792602

Chamero P, Marton TF, Logan DW, Flanagan K, Cruz JR, Saghatelian A, Cravatt BF, Stowers L (2007) Identification of protein pheromones that promote aggressive behaviour. Nature 450(7171):899–902. https://doi.org/10.1038/nature05997

Chiavegatto S, Dawson VL, Mamounas LA, Koliatsos VE, Dawson TM, Nelson RJ (2001) Brain serotonin dysfunction accounts for aggression in male mice lacking neuronal nitric oxide synthase. Proc Natl Acad Sci U S A 98(3):1277–1281. https://doi.org/10.1073/pnas.031487198

Coccaro EF, Sripada CS, Yanowitch RN, Luan Phan K (2011) Corticolimbic function in impulsive aggressive behavior. Biol Psychiatry 69(12):1153–1159. https://doi.org/10.1016/j.biopsych.2011.02.032

Davidson RJ (2000) Dysfunction in the neural circuitry of emotion regulation--a possible prelude to violence. Science 289(5479):591–594. https://doi.org/10.1126/science.289.5479.591

Demas GE, Kriegsfeld LJ, Blackshaw S, Huang P, Gammie SC, Nelson RJ, Snyder SH (1999) Elimination of aggressive behavior in male mice lacking endothelial nitric oxide synthase. J Neurosci Off J Soc Neurosci 19(19):RC30

van Elst T, Ludger BH, Thiel T, Geiger E, Haegele K, Lemieux L, Lieb K, Bohus M, Hennig J, Ebert D

(2003) Frontolimbic brain abnormalities in patients with borderline personality disorder. Biol Psychiatry 54(2):163–171. https://doi.org/10.1016/S0006-3223(02)01743-2

van Erp AM, Miczek KA (2000) Aggressive behavior, increased accumbal dopamine, and decreased cortical serotonin in rats. J Neurosci Off J Soc Neurosci 20 (24):9320–9325

Falkner AL, Dollar P, Perona P, Anderson DJ, Lin D (2014) Decoding ventromedial hypothalamic neural activity during male mouse aggression. J Neurosci Off J Soc Neurosci 34(17):5971–5984. https://doi.org/10.1523/JNEUROSCI.5109-13.2014

Falkner AL, Grosenick L, Davidson TJ, Deisseroth K, Lin D (2016) Hypothalamic control of male aggression-seeking behavior. Nat Neurosci 19(4):596–604. https://doi.org/10.1038/nn.4264

Ferrari PF, Van Erp AMM, Tornatzky W, Miczek KA (2003) Accumbal dopamine and serotonin in anticipation of the next aggressive episode in rats: microdialysis and anticipation of aggression. Eur J Neurosci 17(2):371–378. https://doi.org/10.1046/j.1460-9568.2003.02447.x

Ferris CF (2000) Adolescent stress and neural plasticity in hamsters: a vasopressin-serotonin model of inappropriate aggressive behaviour. Exper Physiol 85 Spec No:85S–90S

Field EF, Whishaw IQ, Pellis SM, Watson NV (2006) Play fighting in androgen-insensitive tfm rats: evidence that androgen receptors are necessary for the development of adult playful attack and defense. Dev Psychobiol 48 (2):111–120. https://doi.org/10.1002/dev.20121

Fortman G, Bas de. (2005) Violence among peoples in the light of human frustration and aggression. Eur J Pharmacol 526(1–3):2–8. https://doi.org/10.1016/j.ejphar.2005.09.035

Frye CA, Walf AA, Kohtz AS, Zhu Y (2014) Progesterone-facilitated lordosis of estradiol-primed mice is attenuated by knocking down expression of membrane progestin receptors in the midbrain. Steroids 81:17–25. https://doi.org/10.1016/j.steroids.2013.11.009

Gammie SC, Nelson RJ (1999) Maternal aggression is reduced in neuronal nitric oxide synthase-deficient mice. J Neurosci Off J Soc Neurosci 19 (18):8027–8035

Gammie SC, Nelson RJ (2001) CFOS and PCREB activation and maternal aggression in mice. Brain Res 898 (2):232–241. https://doi.org/10.1016/s0006-8993(01)02189-8

Gammie SC, Huang PL, Nelson RJ (2000) Maternal aggression in endothelial nitric oxide synthase-deficient mice. Horm Behav 38(1):13–20. https://doi.org/10.1006/hbeh.2000.1595

Giovenardi M, de Azevedo MS, da Silva SP, do E S Hermel E, Gomes CM, Lucion AB (2005) Neonatal handling increases fear and aggression in lactating rats. Physiol Behav 86(1–2):209–217. https://doi.org/10.1016/j.physbeh.2005.07.011

Golden SA, Aleyasin H, Heins R, Flanigan M, Heshmati M, Takahashi A, Russo SJ, Shaham Y (2017a) Persistent conditioned place preference to aggression experience in adult male sexually-experienced CD-1 mice. Genes Brain Behav 16 (1):44–55. https://doi.org/10.1111/gbb.12310

Golden SA, Heins C, Venniro M, Caprioli D, Zhang M, Epstein D, Shaham Y (2017b) Compulsive addiction-like aggressive behavior in mice. Biol Psychiatry 82 (4):239–248. https://doi.org/10.1016/j.biopsych.2017.03.004

Gorlova AV, Pavlov DA, Ushakova VM, Zubkov EA, Morozova AY, Zorkina YA, Inozemtsev AN, Chekhonin VP (2019) Chronic exposure to ultrasonic frequencies selectively increases aggression in rats. Dokl Biol Sci 486(1):69–71. https://doi.org/10.1134/S0012496619030074

Granic I (2014) The role of anxiety in the development, maintenance, and treatment of childhood aggression. Dev Psychopathol 26(4 Pt 2):1515–1530. https://doi.org/10.1017/S0954579414001175

Halász J, Liposits Z, Meelis W, Kruk MR, Haller J (2002) Hypothalamic attack area-mediated activation of the forebrain in aggression. Neuroreport 13 (10):1267–1270. https://doi.org/10.1097/00001756-200207190-00010

Haller J (2017) Studies into abnormal aggression in humans and rodents: methodological and translational aspects. Neurosci Biobehav Rev 76(Pt A):77–86. https://doi.org/10.1016/j.neubiorev.2017.02.022

Haller J (2018a) The role of central and medial amygdala in normal and abnormal aggression: a review of classical approaches. Neurosci Biobehav Rev 85:34–43. https://doi.org/10.1016/j.neubiorev.2017.09.017

Haller J (2018b) Preclinical models of conduct disorder - principles and pharmacologic perspectives. Neurosci Biobehav Rev 91:112–120. https://doi.org/10.1016/j.neubiorev.2016.05.032

Haller J, Kruk MR (2006) Normal and abnormal aggression: human disorders and novel laboratory models. Neurosci Biobehav Rev 30(3):292–303. https://doi.org/10.1016/j.neubiorev.2005.01.005

Harat M, Rudaś M, Zieliński P, Birska J, Sokal P (2015) Deep brain stimulation in pathological aggression. Stereotact Funct Neurosurg 93(5):310–315. https://doi.org/10.1159/000431373

Hasen NS, Gammie SC (2005) Differential Fos activation in virgin and lactating mice in response to an intruder. Physiol Behav 84(5):681–695. https://doi.org/10.1016/j.physbeh.2005.02.010

Hasen NS, Gammie SC (2006) Maternal aggression: new insights from Egr-1. Brain Res 1108(1):147–156. https://doi.org/10.1016/j.brainres.2006.06.007

Hashikawa K, Hashikawa Y, Tremblay R, Zhang J, Feng JE, Sabol A, Piper WT, Lee H, Rudy B, Lin D (2017) Esr1+ cells in the ventromedial hypothalamus control female aggression. Nat Neurosci 20(11):1580–1590. https://doi.org/10.1038/nn.4644

Hernandez-Lallement J, van Wingerden M, Kalenscher T (2018) Towards an animal model of callousness. Neurosci Biobehav Rev 91(August):121–129. https://doi.org/10.1016/j.neubiorev.2016.12.029

Hinsberger M, Sommer J, Kaminer D, Holtzhausen L, Weierstall R, Seedat S, Madikane S, Elbert T (2016) Perpetuating the cycle of violence in south African low-income communities: attraction to violence in young men exposed to continuous threat. Eur J Psychotraumatol 7(January):29099. https://doi.org/10.3402/ejpt.v7.29099

Hong W, Kim D-W, Anderson DJ (2014) Antagonistic control of social versus repetitive self-grooming behaviors by separable amygdala neuronal subsets. Cell 158(6):1348–1361. https://doi.org/10.1016/j.cell.2014.07.049

Hoptman MJ (2015) Impulsivity and aggression in schizophrenia: a neural circuitry perspective with implications for treatment. CNS Spectr 20 (3):280–286. https://doi.org/10.1017/S1092852915000206

Hsu Y, Earley RL, Wolf LL (2006) Modulation of aggressive behaviour by fighting experience: mechanisms and contest outcomes. Biol Rev Camb Philos Soc 81 (1):33–74. https://doi.org/10.1017/S146479310500686X

Huf G, Alexander J, Gandhi P, Allen MH (2016) Haloperidol plus promethazine for psychosis-induced aggression. Cochrane Database Syst Rev 11:CD005146. https://doi.org/10.1002/14651858.CD005146.pub3

de Jong TR, Neumann ID (2018) Oxytocin and aggression. Curr Top Behav Neurosci 35:175–192. https://doi.org/10.1007/7854_2017_13

Juntti SA, Tollkuhn J, Wu MV, Fraser EJ, Soderborg T, Tan S, Honda S-I, Harada N, Shah NM (2010) The androgen receptor governs the execution, but not programming, of male sexual and territorial behaviors. Neuron 66(2):260–272. https://doi.org/10.1016/j.neuron.2010.03.024

Kim-Cohen J, Caspi A, Taylor A, Williams B, Newcombe R, Craig IW, Moffitt TE (2006) MAOA, maltreatment, and gene-environment interaction predicting children's mental health: new evidence and a meta-analysis. Mol Psychiatry 11(10):903–913. https://doi.org/10.1038/sj.mp.4001851

Kisko TM, Wöhr M, Pellis VC, Pellis SM (2017) From play to aggression: high-frequency 50-khz ultrasonic vocalizations as play and appeasement signals in rats. Curr Top Behav Neurosci 30:91–108. https://doi.org/10.1007/7854_2015_432

Knapp P, Chait A, Pappadopulos E, Crystal S, Jensen PS, T-MAY Steering Group (2012) Treatment of maladaptive aggression in youth: CERT guidelines I. engagement, assessment, and management. Pediatrics 129(6):e1562–e1576. https://doi.org/10.1542/peds.2010-1360

Kudryavtseva NN, Smagin DA, Bondar NP (2011) Modeling fighting deprivation effect in mouse repeated aggression paradigm. Prog Neuro-Psychopharmacol Biol Psychiatry 35(6):1472–1478. https://doi.org/10.1016/j.pnpbp.2010.10.013

Lansford JE (2018) Development of aggression. Curr Opin Psychol Aggres Viol 19(February):17–21. https://doi.org/10.1016/j.copsyc.2017.03.015

Laredo SA, Orr VN, McMackin MZ, Trainor BC (2014) The effects of exogenous melatonin and melatonin receptor blockade on aggression and estrogen-dependent gene expression in male California mice (Peromyscus Californicus). Physiol Behav 128 (April):86–91. https://doi.org/10.1016/j.physbeh.2014.01.039

Lee GP, Bechara A, Adolphs R, Arena J, Meador KJ, Loring DW, Smith JR (1998) Clinical and physiological effects of stereotaxic bilateral amygdalotomy for intractable aggression. J Neuropsychiatry Clin Neurosci 10(4):413–420. https://doi.org/10.1176/jnp.10.4.413

Lee H, Kim D-W, Remedios R, Anthony TE, Chang A, Madisen L, Zeng H, Anderson DJ (2014) Scalable control of mounting and attack by Esr1+ neurons in the ventromedial hypothalamus. Nature 509 (7502):627–632. https://doi.org/10.1038/nature13169

Lin D, Boyle MP, Dollar P, Lee H, Lein ES, Perona P, Anderson DJ (2011) Functional identification of an aggression locus in the mouse hypothalamus. Nature 470(7333):221–226. https://doi.org/10.1038/nature09736

Martín-Sánchez A, McLean L, Beynon RJ, Hurst JL, Ayala G, Lanuza E, Martínez-Garcia F (2015) From sexual attraction to maternal aggression: when pheromones change their behavioural significance. Horm Behav 68(February):65–76. https://doi.org/10.1016/j.yhbeh.2014.08.007

McDonald MM, Markham CM, Alisa N, Elliott Albers H, Huhman KL (2012) GABAA receptor activation in the lateral septum reduces the expression of conditioned defeat and increases aggression in Syrian hamsters. Brain Res 1439(February):27–33. https://doi.org/10.1016/j.brainres.2011.12.042

McHenry JA, Robison CL, Bell GA, Bolaños-Guzmán CA, Vialou VV, Nestler EJ, Hull EM (2016) The role of ΔFosB in the medial preoptic area: differential effects of mating and cocaine history. Behav Neurosci 130(5):469–478. https://doi.org/10.1037/bne0000160

Meloni M, Williams S, Martin P (2016) The biosocial: sociological themes and issues. Sociol Rev Monogr 64 (1):7–25. https://doi.org/10.1002/2059-7932.12010

Miczek KA, Maxson SC, Fish EW, Faccidomo S (2001) Aggressive behavioral phenotypes in mice. Behav Brain Res 125(1–2):167–181. https://doi.org/10.1016/s0166-4328(01)00298-4

Miczek KA, Fish EW, Almeida RMM, Faccidomo S, Debold JF (2006) Role of alcohol consumption in escalation to violence. Ann N Y Acad Sci 1036 (1):278–289. https://doi.org/10.1196/annals.1330.018

Miczek KA, Takahashi A, Gobrogge KL, Hwa LS, de Almeida RMM (2015) Escalated aggression in animal models: shedding new light on mesocorticolimbic circuits. Social Behavior 3(June):90–95. https://doi.org/10.1016/j.cobeha.2015.02.007

Mucignat-Caretta C, Cavaggioni A, Caretta A (2004) Male urinary chemosignals differentially affect aggressive behavior in male mice. J Chem Ecol 30(4):777–791

Munley KM, Rendon NM, Demas GE (2018) Neural androgen synthesis and aggression: insights from a seasonally breeding rodent. Front Endocrinol 9:136. https://doi.org/10.3389/fendo.2018.00136

Nakata M, Sano K, Musatov S, Yamaguchi N, Sakamoto T, Ogawa S (2016) Effects of prepubertal or adult site-specific knockdown of estrogen receptor β in the medial preoptic area and medial amygdala on social behaviors in male mice. Eneuro 3(2): ENEURO.0155-15.2016. https://doi.org/10.1523/ENEURO.0155-15.2016

Natarajan D, de Vries H, Saaltink D-J, de Boer SF, Koolhaas JM (2009) Delineation of violence from functional aggression in mice: an ethological approach. Behav Genet 39(1):73–90. https://doi.org/10.1007/s10519-008-9230-3

Nautiyal KM, Tanaka KF, Barr MM, Tritschler L, Le Dantec Y, David DJ, Gardier AM, Blanco C, Hen R, Ahmari SE (2015) Distinct circuits underlie the effects of 5-HT1B receptors on aggression and impulsivity. Neuron 86(3):813–826. https://doi.org/10.1016/j.neuron.2015.03.041

Nelson RJ, Trainor BC (2007) Neural mechanisms of aggression. Nat Rev Neurosci 8(7):536–546. https://doi.org/10.1038/nrn2174

Niederkofler V, Asher TE, Okaty BW, Rood BD, Narayan A, Hwa LS, Beck SG, Miczek KA, Dymecki SM (2016) Identification of serotonergic neuronal modules that affect aggressive behavior. Cell Rep 17(8):1934–1949. https://doi.org/10.1016/j.celrep.2016.10.063

Nyberg J, Sandnabba K, Schalkwyk L, Sluyter F (2004) Genetic and environmental (inter)actions in male mouse lines selected for aggressive and nonaggressive behavior. Genes Brain Behav 3(2):101–109

Olivier B, van Oorschot R (2005) 5-HT1B receptors and aggression: a review. Eur J Pharmacol 526(1–3):207–217. https://doi.org/10.1016/j.ejphar.2005.09.066

Pavlov DA, Gorlova AV, Ushakova VM, Zubkov EA, Morozova AY, Inozemtsev AN, Chekhonin VP (2017) Effects of chronic exposure to ultrasound of alternating frequencies on the levels of aggression and anxiety in CBA and BALB/c mice. Bull Exp Biol Med 163(4):409–411. https://doi.org/10.1007/s10517-017-3815-x

Pompili E, Carlone C, Silvestrini C, Nicolò G (2017) Focus on aggressive behaviour in mental illness. Riv Psichiatr 52(5):175–179. https://doi.org/10.1708/2801.28344

Raine A, Ishikawa SS, Arce E, Lencz T, Knuth KH, Bihrle S, LaCasse L, Colletti P (2004) Hippocampal structural asymmetry in unsuccessful psychopaths. Biol Psychiatry 55(2):185–191. https://doi.org/10.1016/S0006-3223(03)00727-3

Rice CJ, Sandman CA, Lenjavi MR, Baram TZ (2008) A novel mouse model for acute and long-lasting consequences of early life stress. Endocrinology 149(10):4892–4900. https://doi.org/10.1210/en.2008-0633

Robb AS, Schwabe S, Ceresoli-Borroni G, Nasser A, Yu C, Marcus R, Candler SA, Findling RL (2019) A proposed anti-maladaptive aggression agent classification: improving our approach to treating impulsive aggression. Postgrad Med 131(2):129–137. https://doi.org/10.1080/00325481.2019.1574401

Sakurai K, Zhao S, Takatoh J, Rodriguez E, Lu J, Leavitt AD, Min F, Han B-X, Wang F (2016) Capturing and manipulating activated neuronal ensembles with cane delineates a hypothalamic social-fear circuit. Neuron 92(4):739–753. https://doi.org/10.1016/j.neuron.2016.10.015

Seo D, Patrick CJ, Kennealy PJ (2008) Role of serotonin and dopamine system interactions in the neurobiology of impulsive aggression and its comorbidity with other clinical disorders. Aggress Violent Behav 13(5):383–395. https://doi.org/10.1016/j.avb.2008.06.003

Shih JC, Chen K, Ridd MJ (1999) Monoamine oxidase: from genes to behavior. Annu Rev Neurosci 22:197–217. https://doi.org/10.1146/annurev.neuro.22.1.197

Smagin DA, Park J-H, Michurina TV, Peunova N, Glass Z, Sayed K, Bondar NP, Kovalenko IN, Kudryavtseva NN, Enikolopov G (2015) Altered hippocampal neurogenesis and amygdalar neuronal activity in adult mice with repeated experience of aggression. Front Neurosci 9:443. https://doi.org/10.3389/fnins.2015.00443

Spiteri T, Musatov S, Ogawa S, Ribeiro A, Pfaff DW, Agmo A (2010) The role of the estrogen receptor alpha in the medial amygdala and ventromedial nucleus of the hypothalamus in social recognition, anxiety and aggression. Behav Brain Res 210(2):211–220. https://doi.org/10.1016/j.bbr.2010.02.033

Sternson SM (2013) Hypothalamic survival circuits: blueprints for purposive behaviors. Neuron 77(5):810–824. https://doi.org/10.1016/j.neuron.2013.02.018

Stowers L (2002) Loss of sex discrimination and male-male aggression in mice deficient for TRP2. Science 295(5559):1493–1500. https://doi.org/10.1126/science.1069259

Summers CH, Winberg S (2006) Interactions between the neural regulation of stress and aggression. J Exp Biol 209(Pt 23):4581–4589. https://doi.org/10.1242/jeb.02565

Summers CH, Korzan WJ, Lukkes JL, Watt MJ, Forster GL, Øverli Ø, Höglund E et al (2005) Does serotonin influence aggression? Comparing regional activity before and during social interaction. Physiol Biochem Zool 78(5):679–694. https://doi.org/10.1086/432139

Swann AC (2003) Neuroreceptor mechanisms of aggression and its treatment. J Clin Psychiatry 64(Suppl 4):26–35

Takahashi A, Quadros IM, de Almeida RMM, Miczek KA (2012) Behavioral and pharmacogenetics of aggressive behavior. Curr Top Behav Neurosci 12:73–138. https://doi.org/10.1007/7854_2011_191

Takahashi A, Nagayasu K, Nishitani N, Kaneko S, Koide T (2014) Control of intermale aggression by medial prefrontal cortex activation in the mouse. PLoS One 9 (4):e94657. https://doi.org/10.1371/journal.pone.0094657

Tobore TO (2019) On the neurobiological role of oxidative stress in alcohol-induced impulsive, aggressive and suicidal behavior. Substance Use Misuse 54 (14):2290–2303. https://doi.org/10.1080/10826084.2019.1645179

Tóth M, Halász J, Mikics É, Barsy B, Haller J (2008) Early social deprivation induces disturbed social communication and violent aggression in adulthood. Behav Neurosci 122(4):849–854. https://doi.org/10.1037/0735-7044.122.4.849

Toth M, Fuzesi T, Halasz J, Tulogdi A, Haller J (2010) Neural inputs of the hypothalamic 'aggression area' in the rat. Behav Brain Res 215(1):7–20. https://doi.org/10.1016/j.bbr.2010.05.050

Trainor BC, Workman JL, Jessen R, Nelson RJ (2007a) Impaired nitric oxide synthase signaling dissociates social investigation and aggression. Behav Neurosci 121(2):362–369. https://doi.org/10.1037/0735-7044.121.2.362

Trainor BC, Lin S, Sima Finy M, Rowland MR, Nelson RJ (2007b) Photoperiod reverses the effects of estrogens on male aggression via genomic and nongenomic pathways. Proc Natl Acad Sci U S A 104 (23):9840–9845. https://doi.org/10.1073/pnas.0701819104

Tulogdi A, Biro L, Barsvari B, Stankovic M, Haller J, Toth M (2015) Neural mechanisms of predatory aggression in rats—implications for abnormal intraspecific aggression. Behav Brain Res 283(April):108–115. https://doi.org/10.1016/j.bbr.2015.01.030

Unger EK, Burke KJ, Yang CF, Bender KJ, Fuller PM, Shah NM (2015) Medial amygdalar aromatase neurons regulate aggression in both sexes. Cell Rep 10 (4):453–462. https://doi.org/10.1016/j.celrep.2014.12.040

Vasileva O, Balyasnikova N (2019) (Re)Introducing Vygotsky's thought: from historical overview to contemporary psychology. Front Psychol 10:1515. https://doi.org/10.3389/fpsyg.2019.01515

Veenema AH (2009) Early life stress, the development of aggression and neuroendocrine and neurobiological correlates: what can we learn from animal models? Front Neuroendocrinol 30(4):497–518. https://doi.org/10.1016/j.yfrne.2009.03.003

Veenema AH, Neumann ID (2007) Neurobiological mechanisms of aggression and stress coping: a comparative study in mouse and rat selection lines. Brain Behav Evol 70(4):274–285. https://doi.org/10.1159/000105491

Veenema AH, Neumann ID (2009) Maternal separation enhances offensive play-fighting, basal corticosterone and hypothalamic vasopressin MRNA expression in juvenile male rats. Psychoneuroendocrinology 34 (3):463–467. https://doi.org/10.1016/j.psyneuen.2008.10.017

Veenema AH, Torner L, Blume A, Beiderbeck DI, Neumann ID (2007) Low inborn anxiety correlates with high intermale aggression: link to acth response and neuronal activation of the hypothalamic paraventricular nucleus. Horm Behav 51(1):11–19. https://doi.org/10.1016/j.yhbeh.2006.07.004

Veening JG, Coolen LM, de Jong TR, Joosten HW, de Boer SF, Koolhaas JM, Olivier B (2005) Do similar neural systems subserve aggressive and sexual behaviour in male rats? Insights from c-Fos and pharmacological studies. Eur J Pharmacol 526 (1–3):226–239. https://doi.org/10.1016/j.ejphar.2005.09.041

Veit R, Flor H, Erb M, Hermann C, Lotze M, Grodd W, Birbaumer N (2002) Brain circuits involved in emotional learning in antisocial behavior and social phobia in humans. Neurosci Lett 328(3):233–236. https://doi.org/10.1016/S0304-3940(02)00519-0

Vitiello B, Stoff DM (1997) Subtypes of aggression and their relevance to child psychiatry. J Am Acad Child Adolesc Psychiatry 36(3):307–315. https://doi.org/10.1097/00004583-199703000-00008

Wang Y, He Z, Zhao C, Li L (2013) Medial amygdala lesions modify aggressive behavior and immediate early gene expression in oxytocin and vasopressin neurons during intermale exposure. Behav Brain Res 245(May):42–49. https://doi.org/10.1016/j.bbr.2013.02.002

Wang L, Talwar V, Osakada T, Kuang A, Guo Z, Yamaguchi T, Lin D (2019a) Hypothalamic control of conspecific self-defense. Cell Rep 26 (7):1747–1758.e5. https://doi.org/10.1016/j.celrep.2019.01.078

Wang L, Simms J, Peters CJ, Fontaine MT-L, Kexin L, Michael Gill T, Jan YN, Lily Y (2019b) TMEM16B calcium-activated chloride channels regulate action potential firing in lateral septum and aggression in male mice. J Neurosci 39:3137–3118. https://doi.org/10.1523/JNEUROSCI.3137-18.2019

Wong LC, Wang L, D'Amour JA, Yumita T, Chen G, Yamaguchi T, Chang BC et al (2016) Effective modulation of male aggression through lateral septum to medial hypothalamus projection. Curr Biol 26 (5):593–604. https://doi.org/10.1016/j.cub.2015.12.065

Yang Y, Raine A (2009) Prefrontal structural and functional brain imaging findings in antisocial, violent, and psychopathic individuals: a meta-analysis. Psychiatry

Res Neuroimaging 174(2):81–88. https://doi.org/10.1016/j.pscychresns.2009.03.012
Yang CF, Chiang MC, Gray DC, Prabhakaran M, Alvarado M, Juntti SA, Unger EK, Wells JA, Shah NM (2013) Sexually dimorphic neurons in the ventromedial hypothalamus govern mating in both sexes and aggression in males. Cell 153(4):896–909. https://doi.org/10.1016/j.cell.2013.04.017

Neural Regulation of Feeding Behavior

3

Sijia Hao, Yiwen Yang, Mohamed Helmy, and Hao Wang

Abstract

Food intake and energy homeostasis determine survival of the organism and species. Information on total energy levels and metabolic state are sensed in the periphery and transmitted to the brain, where it is integrated and triggers the animal to forage, prey, and consume food. Investigating circuitry and cellular mechanisms coordinating energy balance and feeding behaviors has drawn on many state-of-the-art techniques, including gene manipulation, optogenetics, virus tracing, and single-cell sequencing. These new findings provide novel insights into how the central nervous system regulates food intake, and shed the light on potential therapeutic interventions for eating-related disorders such as obesity and anorexia.

Keywords

Feeding behavior · Energy balance · Hypothalamus · Neural circuits

3.1 Introduction

Feeding determines the survival and reproductive fitness of an organism, and it is a powerful selective pressure in the evolution of a species (Cox et al. 2012). External and internal cues, including food appearance, taste, smell, emotional state, and food preferences together, have a profound impact on feeding regulation. Appropriate feeding decisions are made by integrating sensory information and higher-order brain functions to balance environmental circumstances and internal needs. For example, a long-term shortage of energy resources will reduce the threshold of an animal's risk aversion, and animals are more likely to enter unfamiliar or dangerous territory to find food (Krebs 1980; Magnhagen 1988).

While appetite and food intake are essential for the maintenance of energy homeostasis, in modern human society, easy access to low-cost, high-calorie foods and sedentary lifestyle has dramatically increased the prevalence of obesity and related disorders such as diabetes and cardiovascular disease. Another feeding disorder related to modernization is undereating relating to problematic conditions such as anorexia. Central mechanisms integrate environmental and

---

Sijia Hao and Yiwen Yang contributed equally to this work.

S. Hao · Y. Yang · M. Helmy · H. Wang (✉)
Department of Neurobiology and Department of Neurosurgery of Second Affiliated Hospital, Key Laboratory for Biomedical Engineering of Education Ministry, Zhejiang University School of Medicine, Hangzhou, Zhejiang, China

NHC and CAMS Key Laboratory of Medical Neurobiology, MOE Frontier Science Center for Brain Research and Brain Machine Integration, School of Brain Science and Brain Medicine, Zhejiang University, Hangzhou, Zhejiang, China
e-mail: haowang@zju.edu.cn

H. Wang (ed.), *Neural Circuits of Innate Behaviors*, Advances in Experimental Medicine and Biology 1284, https://doi.org/10.1007/978-981-15-7086-5_3

physiological factors controlling appetite. Investigating the mechanisms underlying neural regulation of feeding behavior is necessary to gain insight into the pathophysiology of feeding disorders.

Appetite, or the motivation to eat, is discussed in the context of either a homeostatic or a hedonic system (Saper et al. 2002). In homeostatic feeding, the animal is in an energy-deficient state. In hedonic feeding, the animal consumes highly palatable food in the absence of an energetic or nutritional need.

Homeostatic feeding occurs in four distinguishable phases: (i) detection of an energy deficit; (ii) goal-directed foraging; (iii) consummatory behavior or food intake; and (iv) termination of feeding. Carrying out of these processes relies on intimate crosstalk between peripheral metabolic organs and the brain. Complex neural networks have evolved to coordinate this essential activity. In this review, we will discuss the molecular and neural circuitry mechanisms of the different stages of homeostatic feeding, as well as hedonic feeding.

3.2 Homeostatic Feeding

3.2.1 Sensing Metabolic State

Homeostatic feeding behavior starts by sensing metabolic state, which is largely dependent on peripheral metabolic organs. Fat tissue, liver, pancreas, gastrointestinal tract (GIT), muscles, and bones are all involved in producing molecular signals that represent body energy state. Upon energy deficit, the appetite-promoting or orexigenic hormone ghrelin is secreted by enteroendocrine cells in the gastrointestinal tract into systemic circulation (Muller et al. 2015). Simultaneously, secretion of insulin from the pancreas and fat-derived hormone leptin is reduced (Ahrén 2000; de Lartigue et al. 2014; Reno et al. 2015). These molecules transmit the information that the body is energy-deficient via blood circulation to the brain, which consequently initiates feeding behavior.

The hypothalamus is a nodal brain structure maintaining energy homeostasis. Specific neuronal subtypes in the hypothalamus detect circulating cues of energy deficit. Neurons expressing agouti-related peptide (AGRP) in the hypothalamic arcuate nucleus (ARC) also express orexigenic peptide neuropeptide Y (NPY) and are GABAergic. These neurons are proximal to the median eminence and are not fully protected by the blood–brain barrier (Yulyaningsih et al. 2017; Schwartz et al. 2000). AGRP neurons directly sense circulating hormones such as leptin (Marco et al. 2004), insulin (KöNner et al. 2007), and ghrelin (Nakazato et al. 2001). Elevated levels of ghrelin in systemic circulation stimulate ARC AGRP/NPY (ARC^{AGRP}) neurons to release GABA, AGRP, and NPY (Kennedy 1950). A broad range of research methods including neural ablation, chemogenetic and optogenetic manipulations of these ARC^{AGRP} neurons showed that stimulation of these neurons triggers feeding behavior while inhibition reduced food intake, even in starved animals (Aponte et al. 2011; Atasoy et al. 2014; Krashes et al. 2013; Krashes et al. 2011). Optogenetic manipulation of ARC^{AGRP} neurons showed that greater recruitment of neurons was associated with larger amounts of food being consumed (Stachniak et al. 2014). Collectively, these results strongly suggest that ARC^{AGRP} neurons work as the "gas pedal" for controlling feeding (Fig. 3.1).

A recent study showed that somatostatin (SST) neurons in the tuberal nucleus (TN) of hypothalamus (TN^{SST}) are directly activated by ghrelin. Ablation of TN^{SST} neurons reduced food intake and weight gain (Luo et al. 2018). There results suggest that in addition to well-characterized ARC^{AGRP} neurons, other neuronal subtypes in the hypothalamus also contribute to sensing an energy deficit.

The hindbrain is also well-recognized as a brain region responsible for processing hunger signals. Vagal afferent fibers from the stomach and intestines sense tension and osmotic pressure. As well as physical signals, nutrients and hormones secreted upon digestion of food convey information from the GIT to the nucleus tractus solitaries (NTS) in hindbrain. NTS also expresses receptors for several hormones and metabolic mediators including leptin (Scott et al. 2009),

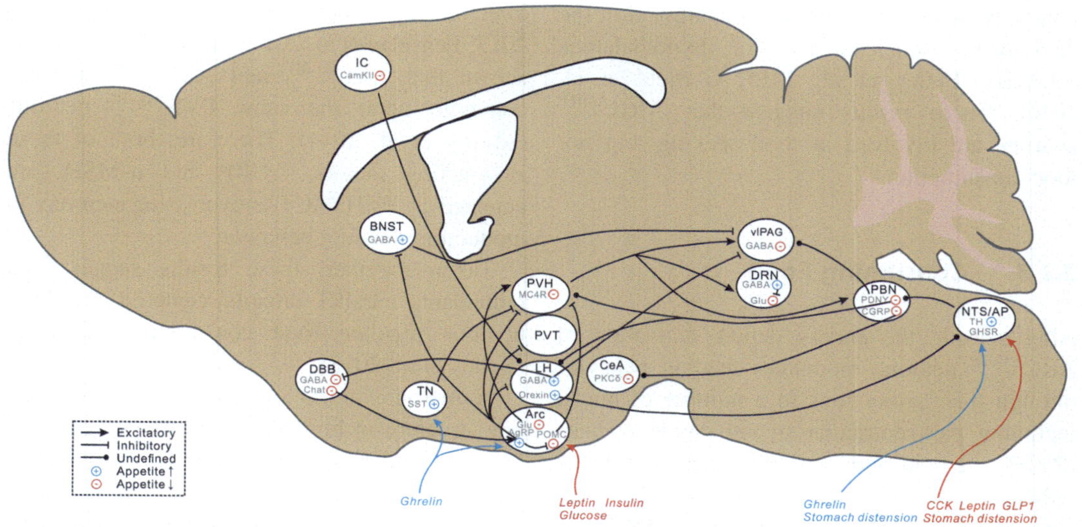

Fig. 3.1 Summary diagram illustrating the neural circuits for homeostatic feeding. *AP* area postrema, *Arc* arcuate hypothalamic nucleus, *BNST* bed nucleus of the stria terminalis, *CeA* central amygdaloid nucleus, *DBB* diagonal band of Broca, *DRN* dorsal raphe nucleus, *IC* insular cortex, *LH* lateral hypothalamus, *NTS* nucleus of the solitary tract, *PBN* parabrachial nucleus, *PVH* paraventricular hypothalamic nucleus, *PVT* paraventricular thalamic nucleus, *TN* tuberal nucleus, *vlPAG* ventrolateral periaqueductal gray

ghrelin (Zigman et al. 2006), and glucose transporter 2 (GLUT2) (Arluison et al. 2004).

Area Postrema (AP) in hindbrain is adjacent to the NTS, and senses circulation signals associated with metabolism such as amylin (Lutz 2013) and leptin (Patterson et al. 2011). The AP sends projections to the NTS. The NTS integrates chemical and electrical inputs that converge upon it and send signals to downstream feeding-related neural networks to start or terminate food consumption.

Bone-derived molecule might also transmit body energy status to the brain and participate in feeding regulation. Lipocalin2 (LCN2) is derived from osteoblasts, crosses the blood–brain barriers, and is a ligand for G-protein subunits α_s-coupled melanocortin 4 receptor (MC4R) (Mosialou et al. 2017), and may therefore modulate feeding through the melanocortin system.

3.2.2 Foraging and Hunting for Food

It is important to reiterate that foraging or hunting for vs. consuming food are two distinguishable phases, though both of them are initiated by a detected energy deficit. That separate circuits mediate these two activities are supported by several lines of evidence. Activation of the GABAergic inputs from the central amygdala (CeA) to ventrolateral periaqueductal gray (vlPAG) in mice induces chasing, pursuing, and killing bite movements, but not consumption of prey (Han et al. 2017). Consistently, during hunting, superior colliculus neurons send out neural signals that are temporally correlated with predatory attacks, but not with feeding after prey capture (Shang et al. 2019).

Additionally, calcium imaging and electrophysiological recording showed that food deprivation augment the activity of ARC^{AGRP} neurons, whereas refeeding quickly reduced the neuronal activity, even before the food is consumed (Betley et al. 2015; Chen et al. 2015; Mandelblat-Cerf et al. 2015). However, sustained inhibition of AGRP neurons requires food consumption. Similarly, evidence from in vivo deep-brain calcium imaging shows that most ARC^{AGRP} neurons activity is reduced upon just the sight of food, presumably through the GABAergic innervations from the ventral compartment of the dorsomedial nucleus of the hypothalamus (vDMH) (Betley et al. 2015; Garfield et al. 2016). These results suggest that ARC^{AGRP} neurons are involved in food seeking, but not food consumption.

3.2.3 Consuming Food

After sensing the energy deficit, activation of ARC^{AGRP} neurons modulates feeding behavior through their projections to a number of nuclei including paraventricular hypothalamic nucleus (PVH), bed nucleus of the stria terminalis (BNST), lateral hypothalamus (LH), and paraventricular thalamus (PVT). ARC^{AGRP} neurons induce feeding behavior in different timescales from minutes to hours through distinct mechanisms. One important mechanism for ARC^{AGRP} neurons mediating fast regulation is through inhibition of the PVH satiety neurons expressing the MC4R (PVH^{MC4R}) by the release of NPY and GABA (Atasoy et al. 2012; Krashes et al. 2013). In a recent study, it was shown that the ARC^{AGRP} neurons could be divided into several subpopulations based on their projection pattern to distinct downstream nuclei in a one-to-one configuration. Among these, activation of the ARC^{AGRP} projections to the anterior BNST, LH, and PVT was sufficient to induce feeding within minutes (Betley et al. 2013).

Slow regulation of energy homeostasis by ARC^{AGRP} neurons is partially attributed to the release of AGRP, which acts as an antagonist to the MC4R in the PVH. In addition, ARC^{AGRP} release GABA which inhibits the anorexigenic proopiomelanocortin (POMC)-expressing neurons in the ARC (ARC^{POMC}). These ARC^{POMC} neurons directly detect anorexigenic molecules such as leptin and insulin in circulation. ARC^{POMC} neurons release alpha melanocyte-stimulating hormone (α-MSH) to reduce food intake and promote weight loss, which plays an opposite role in contrast to ARC^{AGRP} neurons in energy homeostasis. Opto- and chemogenetic activation of ARC^{POMC} neurons inhibits feeding with minor effect and long latency (Aponte et al. 2011; Zhan et al. 2013; Fenselau et al. 2017). Interestingly, studies shown that ARC^{AGRP} and ARC^{POMC} neurons converge upon the same PVH^{MC4R} neurons (Atasoy et al. 2014). Therefore, both of these slow-acting factors, AGRP and α-MSH, are engaged at PVH MC4Rs-expressing neurons in modulating feeding behavior.

Taken together, these results suggest that redundant, parallel circuits centered on ARC neurons regulate food consumption behavior (Betley et al. 2013).

Numerous studies suggest that the PVH is a key structure in homeostatic feeding regulation. Firstly, inhibition of the PVH dramatically increases food intake (Kelly et al. 1979; Atasoy et al. 2012). Lesions of the PVH (Simson et al. 1977; Leibowitz et al. 1981) and haploinsufficiency of SIM1, a transcription factor required for PVH development, cause obesity (Holder Jr. et al. 2000; Michaud et al. 2001). Secondly, silencing PVH^{MC4R} neurons increases appetite (Garfield et al. 2015). Thirdly, activation of the LH^{GABA}-PVH pathway induces feeding behavior (Wu et al. 2015). Finally, studies suggest that PVH^{MC4R} satiety neurons are glutamatergic, and chemo-inhibition of glutamatergic inputs from PVH to dorsal raphe nucleus (DRN) and vlPAG increase food intake, whereas activation of PVH^{MC4R} neurons projections to parabrachial nucleus (PBN) reduce food intake even in starved mice (Garfield et al. 2015). Consistently, DRN neurons are activated by hunger. Stimulation of GABAergic DRN neurons ($DRN^{GABAergic}$) increases food intake, while activation of glutamatergic DRN neurons suppresses food intake. Using single-cell sequencing, specific receptors on DRN neurons were identified for targeting to control food intake (Nectow et al. 2017). GABAergic neurons in the vlPAG also participated, but had an opposite role compared to $DRN^{GABAergic}$ neurons in feeding regulation. Suppression of GABAergic neurons in the vlPAG directly, or through long-projection GABAergic inputs from either BNST or LH, is sufficient to induce feeding behavior quickly in well-fed mice (Hao et al. 2019).

Recent studies revealed a profound role for basal forebrain in appetite control. Cholinergic neurons from the diagonal band of Broca (DBB) in basal forebrain potently suppress food intake through cholinergic inputs to downstream targets in the ARC of hypothalamus (Herman et al. 2016). Additionally, activation of the LH^{GABA}-DBB^{GABA} pathway reduces anxiety and causes indiscriminate feeding (Cassidy et al. 2019).

Two molecularly defined cell subpopulations in the LH, namely orexin/hypocretin-expressing neurons and melanin concentrating hormone (MCH)-producing cells, are important for their role in increasing appetite (Qu et al. 1996; de Lecea et al. 1998; Sakurai et al. 1998). Intracerebroventricular injection of pharmacologic agents showed that both peptides increase food intake and body weight (Qu et al. 1996; Sakurai 1999). Importantly, these neurons not only regulated feeding but also modulated other behaviors such as arousal and sleep (Adamantidis et al. 2007; Jego et al. 2013).

Neurons in hindbrain, specifically NTS, also contribute to feeding behavior. Growth hormone secretagogue receptor (GHSR) is expressed in NTS and other hindbrain areas and is stimulated by ghrelin (Zigman et al. 2006). Direct infusion of ghrelin in NTS/AP increases food intake (Faulconbridge et al. 2003). Orexinergic neurons in LH partially project to tyrosine hydroxylase expressing neurons in A2/C2 region of NTS (Zheng et al. 2005) and their activation increases feeding (Parise et al. 2011). Given that the NTS receives and integrates GIT satiation signals, LH orexinergic projections on it would block the satiation signal, and potentially serve as a disinhibition signal to perpetuate food consumption.

3.2.4 Termination of Feeding

In the process of digestion, several molecules including leptin, amylin, and insulin are released into circulation and suppress appetite (Lutz 2013). Glucagon-like-peptide I (GLP-1) induces satiation and facilitates insulin release into circulation by the pancreas. GLP-1 receptor agonist has been used in the clinical treatment of obesity and diabetes (Finan et al. 2015). Cholecystokinin (CCK) (Fan et al. 2004) and peptide YY (PYY) (Batterham et al. 2007) are also thought to generate satiation. When nutrients are digested by the GIT, enterochromaffin cells release serotonin and stimulate vagal fibers, which facilitates the satiation signal (Alcaino et al. 2018). Distension of the GIT also contributes to inhibition of feeding by sending visceral inputs to the brain (Eisen et al. 2001).

Elevated circulation levels of leptin and insulin after a meal directly activate ARC^{POMC} neurons (Poggioli et al. 1986). Activity of ARC^{POMC} neurons increases onefold after standard diet, but not after high-fat diet, suggesting their specific role in homeostatic feeding. Astrocytes surrounding ARC^{POMC} retract during hyperglycemia, which makes ARC^{POMC} neurons more exposed to satiation circulation signals (Nuzzaci et al. 2020). Disinhibition through silencing of GABAergic ARC^{AGRP} inputs is another factor that augments ARC^{POMC} neuronal activity. Satiation signals are transmitted to downstream brain regions such as PVH and act through the MC4R system to terminate feeding (Huszar et al. 1997).

A subpopulation of glutamate-releasing neurons, which co-express leptin and oxytocin receptors, is found in the ARC (ARC^{Glut}). The projections of these ARC^{Glut} neurons converge with GABAergic ARC^{AGRP} projections to form synapses on PVH^{MC4R} neurons. Chemo- or optogenetic excitation of the ARC^{Glut} to PVH^{MC4R} pathway rapidly caused satiety and reduced food intake. Therefore, ARC^{Glut} neurons provide fast satiety control of feeding (Blevins and Ho 2013; Blevins and Baskin 2015; Fenselau et al. 2017).

The PBN, a nucleus that relays taste and visceral sensory information from NTS, has been shown to be essential for producing satiety and terminating food intake. After sensing metabolic mediators, the NTS transmits signals via glutamatergic innervations to the PBN. Optogenetic activation of a subpopulation of excitatory neurons in the PBN that expresses calcitonin gene-related peptide (PBN^{CGRP}) dramatically reduced food intake. The PBN^{CGRP} neurons project to the laterocapsular division of the CeA forming a functionally important circuit for suppressing appetite (Carter et al. 2013).

As mentioned above, in addition to circulating metabolic mediators, mechanosensory signaling can also inhibit feeding. Neurons in PBN that express prodynorphin gene (PBN^{pdyn}) receive inputs from trigeminal nuclei and subregions of NTS, which receive oral, oropharyngeal, and visceral sensory information. PBN^{pdyn} neurons respond to mechanical pressure in digestive tract which is caused by drinking, eating, or other physical stimuli-like tongue touching or gavage needle insertion into esophagus. Stimulation of PBN^{pdyn} reduced appetite by inhibiting eating bout initiation, but did not terminate eating after the bout is already in progress. Several nuclei in LH, PVH, and vlPAG involved in feeding behavior are innervated by PBN^{pdyn}, forming a feedback control circuit which responds to mechanosensory information from the GIT (Kim et al. 2020).

Higher cortical areas are also involved in processing satiation signals. Neuronal activity in dorso-lateral prefrontal cortex (dlPFC) increased during food consumption as satiation was approached. Compared to lean individuals, obese individuals showed a lower increase in dlPFC neuron activity after digestion (Gluck et al. 2017).

Different neuronal populations in insular cortex were selectively activated or inhibited by food cues during hunger. These neurons no longer responded to a food cue when animals were in a state of satiety (Yamamoto 1984; Livneh et al. 2017; Livneh et al. 2020). The calcium/calmodulin-dependent protein kinase II (CAMKII)-expressing neurons in right anterior insular cortex send projections to LH to suppress feeding in the presence of aversive visceral stimuli (Wu et al. 2020).

3.3 Hedonic Feeding

Beyond metabolic needs, animals prefer to eat more when presented with highly palatable food. This is because palatable food is rich in energy and activates the brain reward system. High-energy food can stimulate reward brain centers starting from taste, which is relayed by the NTS to LH via PBN (Norgren 1974; Moga et al. 1990).

The LH is involved in homeostatic feeding regulation as discussed above, and together with nucleus accumbens (NAc) and ventral tegmental area (VTA), these areas form a reward circuit for hedonic feeding (Fig. 3.2). Neurons in the LH are composed of a number of genetically and functionally distinct cell types. GABAergic and glutamatergic neurons play opposing functions in feeding and reward. Optogenetic activation of VGAT-expressing LH neurons causes voracious feeding, as well as optical self-stimulation behavior (Jennings et al. 2015), whereas activation of LH Vglut2-expressing neurons reduces food intake and produce aversive responses (Jennings et al. 2013; Li et al. 2018). Similarly, a recent study combining single-cell RNA sequencing and in vivo two-photon calcium imaging showed that the transcriptional profile of LH glutamatergic neurons was affected by obesity. Encoding properties of individual LH glutamatergic neurons showed greatly attenuated reward responses in obese mice (Rossi et al. 2019).

GABAergic inputs from ventral BNST preferentially target LH glutamatergic neurons. Activation of this $vBNST^{GABA}$-LH pathway evoke voracious feeding of highly palatable food and induce self-stimulation (Jennings et al. 2015). Another major input onto LH is from NAc shell medium spiny principal neurons (MSNs) which express D1R. D1R-expressing MSNs preferentially innervate LH GABAergic neurons, and not orexin- or MCH-producing neurons. Optogenetic stimulation of the NAc $shell^{D1R}$-LH^{GABA} pathway suppresses licking for a palatable reward, whereas optogenetic inhibition of postsynaptic LH GABAergic neurons suppresses consumption of food (O'Connor et al. 2015). Another subpopulation of NAc neurons which express D2R project to the ventral pallidum, and activation of this pathway is involved in processing taste palatability (Smith and Berridge 2007), and increases feeding via the LH (Stratford and Wirtshafter 2013).

LH^{GABA} neurons directly project to the VTA. Activation of this pathway induces feeding, and

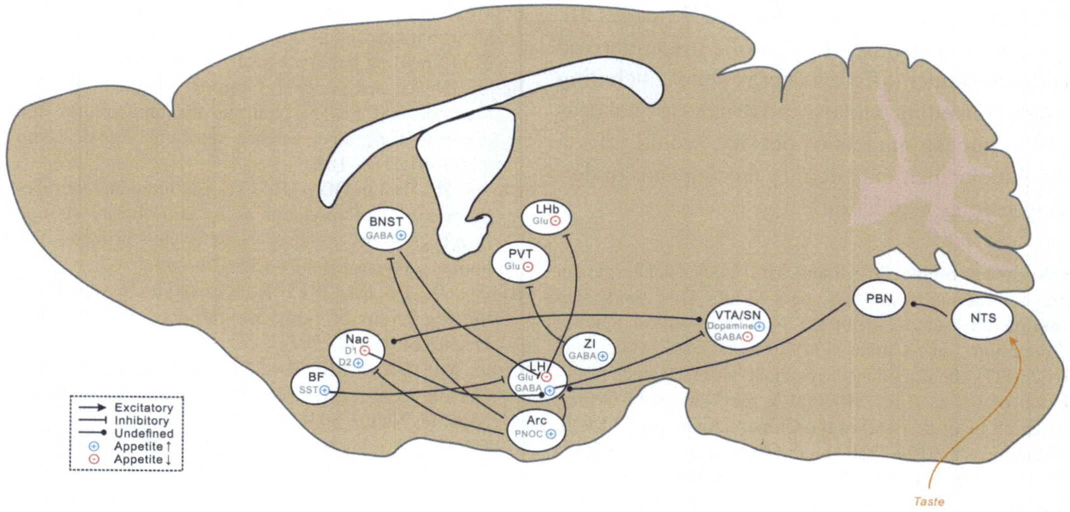

Fig. 3.2 Summary diagram showing the neural circuits for hedonic feeding. *Arc* arcuate hypothalamic nucleus, *BF* basal forebrain, *BNST* bed nucleus of the stria terminalis, *LH* lateral hypothalamus, *LHb* lateral habenular nucleus, *Nac* nucleus accumben, *NTS* nucleus of the solitary tract, *PBN* parabrachial nucleus, *PVT* paraventricular thalamic nucleus, *SN* substantia nigra, *VTA* ventral tegmental area, *ZI* zona incerta

mice will readily engage in optical self-stimulation of this pathway (Nieh et al. 2015). LH^{Vglut2} neurons have a pronounced excitatory projection to the lateral habenula (LHb), a brain area known to mediate the negative emotional valence (Matsumoto and Hikosaka 2007; Stamatakis and Stuber 2012). Optogenetic inhibition of LH^{Vglut2}-LHb projections selectively increases licking for a caloric reward and causes place preference (Stamatakis et al. 2016).

A couple of recent literatures has linked some new nuclei with hedonic feeding. Optogenetic stimulation of GABAergic cells in the zona incerta or their projections to PVT rapidly induced binge-like eating and showed positive hedonic association (Zhang and van den Pol 2017). Optogenetic activation of basal forebrain SST neurons or their projection to the LH specifically leads to high-calorie food intake but not normal chow, and induces anxiety-like behaviors. This study indicated a selective role of basal forebrain SOM neurons in hedonic feeding (Zhu et al. 2017). A subpopulation of GABAergic cells in the ARC, prepronociceptin-expressing neurons (ARC^{PNOC}), were found to be activated by a short-term high-fat diet. The ARC^{PNOC} neurons provide inhibitory synaptic input to nearby POMC neurons and BNST to regulate hedonic feeding. ARC^{PNOC} neurons selectively increase consumption of palatable food, but not normal chow (Jais et al. 2020).

3.4 Conclusion

Recent advances in neuroscience methods have greatly advanced our understanding of neural circuits of feeding behavior. Homeostatic and hedonic feeding have distinct circuitry mechanisms, and share some nodal structures such as the ARC and LH. However, considering the multitude of heterogeneous of neuronal subtypes and spatiotemporal activity dynamics of neurons, a detailed map the neural correlates of feeding behavior is far from complete. High-throughput single-cell transcriptional profiling is a powerful technique that may greatly accelerate our efforts toward mapping central systems coordinating feeding. How the brain interacts with peripheral organs is an emerging and exciting area of research. More comprehensive integration studies of neurobiology, neuroendocrinology, and metabology would enrich our understanding of the brain–body interface in feeding (Levinthal and Strick 2020). Systematic screening of metabolic organ-derived mediators under different energy states, including fasting, refeeding, and excessive calorie intake, as well as mouse models of obesity, would inform novel molecular and circuitry mechanisms underlying feeding regulation.

Acknowledgments We thank Mrs. Lucy McDougle for critical comments on this manuscript. This work was supported by grants from the National Natural Science Foundation of China (31970940, 31671100, and 31622027), the Zhejiang Provincial Natural Science Foundation of China (LR18H090001), the Non-profit Central Research Institute Fund of the Chinese Academy of Medical Sciences (2018PT31041), the Program for Introducing Talents in Discipline to Universities, and the Fundamental Research Funds for the Central Universities (2019QNA5001).

References

Adamantidis AR, Zhang F, Aravanis AM, Deisseroth K, de Lecea L (2007) Neural substrates of awakening probed with optogenetic control of hypocretin neurons. Nature 450(7168):420–424

Ahrén B (2000) Autonomic regulation of islet hormone secretion--implications for health and disease. Diabetologia 43(4):393–410

Alcaino C, Knutson KR, Treichel AJ, Yildiz G, Strege PR, Linden DR, Li JH, Leiter AB, Szurszewski JH, Farrugia G (2018) A population of gut epithelial enterochromaffin cells is mechanosensitive and requires Piezo2 to convert force into serotonin release. Proc Natl Acad Sci 115(32):E7632–E7641

Aponte Y, Atasoy D, Sternson SM (2011) AGRP neurons are sufficient to orchestrate feeding behavior rapidly and without training. Nature Neurosci 14(3):351–355

Arluison M, Quignon M, Nguyen P, Thorens B, Leloup C, Penicaud L (2004) Distribution and anatomical localization of the glucose transporter 2 (GLUT2) in the adult rat brain—an immunohistochemical study. J Chem Neuroanat 28(3):117–136

Atasoy D, Betley JN, Su HH, Sternson SM (2012) Deconstruction of a neural circuit for hunger. Nature 488 (7410):172–177

Atasoy D, Betley JN, Li WP, Su HH, Sertel SM, Scheffer LK, Simpson JH, Fetter RD, Sternson SM (2014) A genetically specified connectomics approach applied to long-range feeding regulatory circuits. Nat Neurosci 17 (12):1830–1839

Batterham RL, Rosenthal JM, Zelaya FO, Barker GJ, Withers DJ, Williams SC (2007) PYY modulation of cortical and hypothalamic brain areas predicts feeding behaviour in humans. Nature 450(7166):106–109

Betley JN, Cao ZF, Ritola KD, Sternson SM (2013) Parallel, redundant circuit organization for homeostatic control of feeding behavior. Cell 155(6):1337–1350

Betley JN, Xu S, Cao ZFH, Gong R, Magnus CJ, Yu Y, Sternson SM (2015) Neurons for hunger and thirst transmit a negative-valence teaching signal. Nature 521(7551):180–185

Blevins JE, Baskin DG (2015) Translational and therapeutic potential of oxytocin as an anti-obesity strategy: insights from rodents, nonhuman primates and humans. Physiol Behav 152:438–449

Blevins JE, Ho JM (2013) Role of oxytocin signaling in the regulation of body weight. Rev Endocr Metab Disord 14(4):311–329

Carter ME, Soden ME, Zweifel LS, Palmiter RD (2013) Genetic identification of a neural circuit that suppresses appetite. Nature 503(7474):111

Cassidy RM, Lu Y, Jere M, Tian JB, Xu Y, Mangieri LR, Felix-Okoroji B, Selever J, Xu Y, Arenkiel BR, Tong Q (2019) A lateral hypothalamus to basal forebrain neurocircuit promotes feeding by suppressing responses to anxiogenic environmental cues. Sci Adv 5(3):eaav1640

Chen Y, Lin YC, Kuo TW, Knight ZA (2015) Sensory detection of food rapidly modulates arcuate feeding circuits. Cell 160(5):829–841

Cox PG, Rayfield EJ, Fagan MJ, Herrel A, Pataky TC, Jeffery N (2012) Functional evolution of the feeding system in rodents. PLoS One 7(4):e36299–e36299

Eisen S, Davis JD, Rauhofer E, Smith GP (2001) Gastric negative feedback produced by volume and nutrient during a meal in rats. Am J Phys Regul Integr Comp Phys 281(4):R1201

Fan W, Ellacott KL, Halatchev IG, Takahashi K, Yu P, Cone RD (2004) Cholecystokinin-mediated suppression of feeding involves the brainstem melanocortin system. Nat Neurosci 7(4):335–336

Faulconbridge LF, Cummings DE, Kaplan JM, Grill HJ (2003) Hyperphagic effects of brainstem ghrelin administration. Diabetes 52(9):2260–2265

Fenselau H, Campbell JN, Verstegen AMJ, Madara JC, Xu J, Shah BP, Resch JM, Yang ZF, Mandelblat-Cerf Y, Livneh Y, Lowell BB (2017) A rapidly acting glutamatergic ARC -> PVH satiety circuit postsynaptically regulated by alpha-MSH (vol 20, pg 42, 2017). Nat Neurosci 20(8):1189–1189

Finan B, Clemmensen C, Müller TD (2015) Emerging opportunities for the treatment of metabolic diseases: Glucagon-like peptide-1 based multi-agonists. Mol Cell Endocrinol 418:42–54

Garfield AS, Li C, Madara JC, Shah BP, Webber E, Steger JS, Campbell JN, Gavrilova O, Lee CE, Olson DP, Elmquist JK, Tannous BA, Krashes MJ, Lowell BB (2015) A neural basis for melanocortin-4 receptor-regulated appetite. Nat Neurosci 18(6):863–871

Garfield AS, Shah BP, Burgess CR, Li MM, Li C, Steger JS, Madara JC, Campbell JN, Kroeger D, Scammell TE, Tannous BA, Myers MG Jr, Andermann ML, Krashes MJ, Lowell BB (2016) Dynamic GABAergic

afferent modulation of AgRP neurons. Nat Neurosci 19 (12):1628–1635
Gluck ME, Viswanath P, Stinson EJ (2017) Obesity, appetite, and the prefrontal cortex. Curr Obes Rep 6 (4):380–388
Han W, Tellez LA, Rangel MJ Jr, Motta SC, Zhang X, Perez IO, Canteras NS, Shammah-Lagnado SJ, van den Pol AN, de Araujo IE (2017) Integrated control of predatory hunting by the central nucleus of the amygdala. Cell 168(1–2):311–324.e318
Hao S, Yang H, Wang X, He Y, Xu H, Wu X, Pan L, Liu Y, Lou H, Xu H, Ma H, Xi W, Zhou Y, Duan S, Wang H (2019) The lateral hypothalamic and BNST GABAergic projections to the anterior ventrolateral periaqueductal Gray regulate feeding. Cell Rep 28 (3):616–624. e615
Herman AM, Ortiz-Guzman J, Kochukov M, Herman I, Quast KB, Patel JM, Tepe B, Carlson JC, Ung K, Selever J, Tong Q, Arenkiel BR (2016) A cholinergic basal forebrain feeding circuit modulates appetite suppression. Nature 538(7624):253–256
Holder JL Jr, Butte NF, Zinn AR (2000) Profound obesity associated with a balanced translocation that disrupts the SIM1 gene. Hum Mol Genet 9(1):101–108
Huszar D, Lynch CA, Fairchild-Huntress V, Dunmore JH, Fang Q, Berkemeier LR, Gu W, Kesterson RA, Boston BA, Cone RD, Smith FJ, Campfield LA, Burn P, Lee F (1997) Targeted disruption of the melanocortin-4 receptor results in obesity in mice. Cell 88(1):131–141
Jais A, Paeger L, Sotelo-Hitschfeld T, Bremser S, Prinzensteiner M, Klemm P, Mykytiuk V, Widdershooven PJM, Vesting AJ, Grzelka K, Minère M, Cremer AL, Xu J, Korotkova T, Lowell BB, Zeilhofer HU, Backes H, Fenselau H, Wunderlich FT, Kloppenburg P, Brüning JC (2020) PNOC(ARC) neurons promote hyperphagia and obesity upon high-fat-diet feeding. Neuron 106:1009
Jego S, Glasgow SD, Herrera CG, Ekstrand M, Reed SJ, Boyce R, Friedman J, Burdakov D, Adamantidis AR (2013) Optogenetic identification of a rapid eye movement sleep modulatory circuit in the hypothalamus. Nat Neurosci 16(11):1637–1643
Jennings JH, Rizzi G, Stamatakis AM, Ung RL, Stuber GD (2013) The inhibitory circuit architecture of the lateral hypothalamus orchestrates feeding. Science 341 (6153):1517–1521
Jennings JH, Ung RL, Resendez SL, Stamatakis AM, Taylor JG, Huang J, Veleta K, Kantak PA, Aita M, Shilling-Scrivo K, Ramakrishnan C, Deisseroth K, Otte S, Stuber GD (2015) Visualizing hypothalamic network dynamics for appetitive and consummatory behaviors. Cell 160(3):516–527
Kelly J, Rothstein J, Grossman SP (1979) GABA and hypothalamic feeding systems. I. Topographic analysis of the effects of microinjections of muscimol. Physiol Behav 23(6):1123–1134
Kennedy GC (1950) The hypothalamic control of food intake in rats. Proc R Soc Lond B Biol Sci 137 (889):535–549
Kim D-Y, Heo G, Kim M, Kim H, Jin JA, Kim H-K, Jung S, An M, Ahn BH, Park JH, Park H-E, Lee M, Lee JW, Schwartz GJ, Kim S-Y (2020) A neural circuit mechanism for mechanosensory feedback control of ingestion. Nature 580(7803):376–380
KöNner AC, Janoschek R, Plum L, Jordan SD, Rother E, Ma X, Xu C, Enriori P, Hampel B, Barsh GS (2007) Insulin action in AgRP-expressing neurons is required for suppression of hepatic glucose production. Cell Metabol 5(6):438–449
Krashes MJ, Koda S, Ye CP, Rogan SC, Adams AC, Cusher DS, Maratos-Flier E, Roth BL, Lowell BB, Koda S (2011) Rapid, reversible activation of AgRP neurons drives feeding behavior in mice. J Clin Investig 121(4):1424–1428
Krashes MJ, Shah BP, Koda S, Lowell BB (2013) Rapid versus delayed stimulation of feeding by the endogenously released agrp neuron mediators GABA, NPY, and AgRP. Cell Metabol 18(4):588–595
Krebs JR (1980) Optimal foraging, predation risk and territory defense. Ardea 68:83–90
de Lartigue G, Ronveaux CC, Raybould HE (2014) Deletion of leptin signaling in vagal afferent neurons results in hyperphagia and obesity. Mol Metab 3(6):595–607
de Lecea L, Kilduff TS, Peyron C, Gao X, Foye PE, Danielson PE, Fukuhara C, Battenberg EL, Gautvik VT, Bartlett FS 2nd, Frankel WN, van den Pol AN, Bloom FE, Gautvik KM, Sutcliffe JG (1998) The hypocretins: hypothalamus-specific peptides with neuroexcitatory activity. Proc Natl Acad Sci U S A 95(1):322–327
Leibowitz SF, Hammer NJ, Chang K (1981) Hypothalamic paraventricular nucleus lesions produce overeating and obesity in the rat. Physiol Behav 27 (6):1031–1040
Levinthal DJ, Strick PL (2020) Multiple areas of the cerebral cortex influence the stomach. Proc Natl Acad Sci USA 117(23):13078–13083
Li Y, Zeng J, Zhang J, Yue C, Zhong W, Liu Z, Feng Q, Luo M (2018) Hypothalamic circuits for predation and evasion. Neuron 97(4):911–924.e915
Livneh Y, Ramesh RN, Burgess CR, Levandowski KM, Madara JC, Fenselau H, Goldey GJ, Diaz VE, Jikomes N, Resch JM, Lowell BB, Andermann ML (2017) Homeostatic circuits selectively gate food cue responses in insular cortex. Nature 546(7660):611–616
Livneh Y, Sugden AU, Madara JC, Essner RA, Flores VI, Sugden LA, Resch JM, Lowell BB, Andermann ML (2020) Estimation of current and future physiological states in insular cortex. Neuron 105(6):1094–1111.e1010
Luo SX, Huang J, Li Q, Mohammad H, Lee CY, Krishna K, Kok AM, Tan YL, Lim JY, Li H, Yeow LY, Sun J, He M, Grandjean J, Sajikumar S, Han W, Fu Y (2018) Regulation of feeding by somatostatin neurons in the tuberal nucleus. Science 361 (6397):76–81
Lutz TA (2013) The interaction of amylin with other hormones in the control of eating. Diabetes Obes Metab 15(2):99–111
Magnhagen C (1988) Predation risk and foraging in juvenile pink (Oncorhynchus-Gorbuscha) and chum Salmon (O-Keta). Can J Fish Aquat Sci 45(4):592–596

Mandelblat-Cerf Y, Ramesh RN, Burgess CR, Patella P, Yang Z, Lowell BB, Andermann ML (2015) Arcuate hypothalamic AgRP and putative POMC neurons show opposite changes in spiking across multiple timescales. elife 4:e07122

Marco VDT, Lee K, Whyment AD, Blanks AM, Spanswick D (2004) Orexigen-sensitive NPY/AgRP pacemaker neurons in the hypothalamic arcuate nucleus. Nat Neurosci 7(5):493–494

Matsumoto M, Hikosaka O (2007) Lateral habenula as a source of negative reward signals in dopamine neurons. Nature 447(7148):1111–1115

Michaud JL, Boucher F, Melnyk A, Gauthier F, Goshu E, Lévy E, Mitchell GA, Himms-Hagen J, Fan CM (2001) Sim1 haploinsufficiency causes hyperphagia, obesity and reduction of the paraventricular nucleus of the hypothalamus. Hum Mol Genet 10 (14):1465–1473

Moga MM, Herbert H, Hurley KM, Yasui Y, Gray TS, Saper CB (1990) Organization of cortical, basal forebrain, and hypothalamic afferents to the parabrachial nucleus in the rat. J Comp Neurol 295(4):624–661

Mosialou I, Shikhel S, Liu J-M, Maurizi A, Luo N, He Z, Huang Y, Zong H, Friedman RA, Barasch J (2017) MC4R-dependent suppression of appetite by bone-derived lipocalin 2. Nature 543(7645):385–390

Muller T, Nogueiras R, Andermann M, Andrews Z, Anker S, Argente J, Batterham R, Benoit S, Bowers C, Broglio F (2015) Ghrelin. Mol Metabol 4 (6):437–460

Nakazato M, Murakami N, Date Y, Kojima M, Matsuo H, Kangawa K, Matsukura S (2001) A role for ghrelin in the central regulation of feeding. Nature 409 (6817):194–198

Nectow AR, Schneeberger M, Zhang HX, Field BC, Renier N, Azevedo E, Patel B, Liang YP, Mitra S, Tessier-Lavigne M, Han MH, Friedman JM (2017) Identification of a brainstem circuit controlling feeding. Cell 170(3):429

Nieh EH, Matthews GA, Allsop SA, Presbrey KN, Leppla CA, Wichmann R, Neve R, Wildes CP, Tye KM (2015) Decoding neural circuits that control compulsive sucrose seeking. Cell 160(3):528–541

Norgren R (1974) Gustatory afferents to ventral forebrain. Brain Res 81(2):285–295

Nuzzaci D, Cansell C, Lienard F, Nedelec E, Ben Fradj S, Castel J, Foppen E, Denis R, Grouselle D, Laderriere A, Lemoine A, Mathou A, Tolle V, Heurtaux T, Fioramonti X, Audinat E, Penicaud L, Nahon JL, Rovere C, Benani A (2020) Postprandial hyperglycemia stimulates neuroglial plasticity in hypothalamic pomc neurons after a balanced meal. Cell Rep 30(9):3067–3078.e3065

O'Connor EC, Kremer Y, Lefort S, Harada M, Pascoli V, Rohner C, Lüscher C (2015) Accumbal D1R neurons projecting to lateral hypothalamus authorize feeding. Neuron 88(3):553–564

Parise EM, Lilly N, Kay K, Dossat AM, Seth R, Overton JM, Williams DL (2011) Evidence for the role of hindbrain orexin-1 receptors in the control of meal size. Am J Physiol Regul Integr Comp Physiol 301 (6):R1692–R1699

Patterson CM, Leshan RL, Jones JC, Myers MG Jr (2011) Molecular mapping of mouse brain regions innervated by leptin receptor-expressing cells. Brain Res 1378:18–28

Poggioli R, Vergoni AV, Bertolini A (1986) ACTH-(1–24) and α-MSH antagonize feeding behavior stimulated by kappa opiate agonists. Peptides 7 (5):843–848

Qu D, Ludwig DS, Gammeltoft S, Piper M, Pelleymounter MA, Cullen MJ, Mathes WF, Przypek R, Kanarek R, Maratos-Flier E (1996) A role for melanin-concentrating hormone in the central regulation of feeding behaviour. Nature 380(6571):243–247

Reno CM, Ding Y, Sherwin R (2015) Leptin acts in the brain to influence hypoglycemic counterregulation: disparate effects of acute and recurrent hypoglycemia on glucagon release. Am J Physiol Endocrinol Metabol 309(12):E960–E967

Rossi MA, Basiri ML, McHenry JA, Kosyk O, Otis JM, van den Munkhof HE, Bryois J, Hübel C, Breen G, Guo W, Bulik CM, Sullivan PF, Stuber GD (2019) Obesity remodels activity and transcriptional state of a lateral hypothalamic brake on feeding. Science 364 (6447):1271–1274

Sakurai T (1999) Orexins and orexin receptors: implication in feeding behavior. Regul Pept 85(1):25–30

Sakurai T, Amemiya A, Ishii M, Matsuzaki I, Chemelli RM, Tanaka H, Williams SC, Richardson JA, Kozlowski GP, Wilson S, Arch JR, Buckingham RE, Haynes AC, Carr SA, Annan RS, McNulty DE, Liu WS, Terrett JA, Elshourbagy NA, Bergsma DJ, Yanagisawa M (1998) Orexins and orexin receptors: a family of hypothalamic neuropeptides and G protein-coupled receptors that regulate feeding behavior. Cell 92(4):573–585

Saper CB, Chou TC, Elmquist JK (2002) The need to feed: homeostatic and hedonic control of eating. Neuron 36 (2):199–211

Schwartz MW, Woods SC, Porte D Jr, Seeley RJ, Baskin DG (2000) Central nervous system control of food intake. Nature 404(6778):661–671

Scott MM, Lachey JL, Sternson SM, Lee CE, Elias CF, Friedman JM, Elmquist JK (2009) Leptin targets in the mouse brain. J Comp Neurol 514(5):518–532

Shang C, Liu A, Li D, Xie Z, Chen Z, Huang M, Li Y, Wang Y, Shen WL, Cao P (2019) A subcortical excitatory circuit for sensory-triggered predatory hunting in mice. Nat Neurosci 22(6):909–920

Simson EL, Gold RM, Standish LJ, Pellett PL (1977) Axon-sparing brain lesioning technique: the use of monosodium-L-glutamate and other amino acids. Science 198(4316):515–517

Smith KS, Berridge KC (2007) Opioid limbic circuit for reward: interaction between hedonic hotspots of nucleus accumbens and ventral pallidum. J Neurosci 27(7):1594–1605

Stachniak TJ, Ghosh A, Sternson SM (2014) Chemogenetic synaptic silencing of neural circuits localizes a hypothalamus→midbrain pathway for feeding behavior. Neuron 82(4):797–808

Stamatakis AM, Stuber GD (2012) Activation of lateral habenula inputs to the ventral midbrain promotes behavioral avoidance. Nat Neurosci 15(8):1105–1107

Stamatakis AM, Van Swieten M, Basiri ML, Blair GA, Kantak P, Stuber GD (2016) Lateral hypothalamic area Glutamatergic neurons and their projections to the lateral Habenula regulate feeding and reward. J Neurosci 36(2):302–311

Stratford TR, Wirtshafter D (2013) Lateral hypothalamic involvement in feeding elicited from the ventral pallidum. Eur J Neurosci 37(4):648–653

Wu Z, Kim ER, Sun H, Xu Y, Mangieri LR, Li DP, Pan HL, Xu Y, Arenkiel BR, Tong Q (2015) GABAergic projections from lateral hypothalamus to paraventricular hypothalamic nucleus promote feeding. J Neurosci 35(8):3312–3318

Wu Y, Chen C, Chen M, Qian K, Lv X, Wang H, Jiang L, Yu L, Zhuo M, Qiu S (2020) The anterior insular cortex unilaterally controls feeding in response to aversive visceral stimuli in mice. Nat Commun 11(1):640

Yamamoto T (1984) Gustatory responses of cortical neurons in rats. I. Response characteristics. J Neurophysiol 51:616

Yulyaningsih E, Rudenko IA, Valdearcos M, Dahlén E, Vagena E, Chan A, Alvarez-Buylla A, Vaisse C, Koliwad SK, Xu AW (2017) Acute lesioning and rapid repair of hypothalamic neurons outside the blood-brain barrier. Cell Rep 19(11):2257–2271

Zhan C, Zhou J, Feng Q, Zhang JE, Lin S, Bao J, Wu P, Luo M (2013) Acute and long-term suppression of feeding behavior by POMC neurons in the brainstem and hypothalamus, respectively. J Neurosci 33 (8):3624–3632

Zhang X, van den Pol AN (2017) Rapid binge-like eating and body weight gain driven by zona incerta GABA neuron activation. Science 356(6340):853–859

Zheng H, Patterson LM, Berthoud HR (2005) Orexin-a projections to the caudal medulla and orexin-induced c-Fos expression, food intake, and autonomic function. J Comp Neurol 485(2):127–142

Zhu C, Yao Y, Xiong Y, Cheng M, Chen J, Zhao R, Liao F, Shi R, Song S (2017) Somatostatin neurons in the basal forebrain promote high-calorie food intake. Cell Rep 20(1):112–123

Zigman JM, Jones JE, Lee CE, Saper CB, Elmquist JK (2006) Expression of ghrelin receptor mRNA in the rat and the mouse brain. J Comp Neurol 494(3):528

Neural Circuits for Reward

4

Wen Zhang

Abstract

Innate behaviors often viewed as genetically predetermined behaviors. However, in the environment animals often are subjected to external stimuli conflicting with those. Thus, animals subsequently need to change those behaviors to survive and reproduce. In the brain, the reward pathway is well-known for its role to adjust behaviors according to external stimuli, or rewards. However, only recently the relationship between reward pathway and innate behavior begins to be explored. In this review, we summarize the recent data on this subject from rodent studies which suggest an important role of this crosstalk between circuits involved in reward pathway and innate behaviors. We also discuss some of the neurotransmitters and neuromodulators underlying this crosstalk and the related mechanisms.

Keywords

Innate behavior · Reward pathway · Dopamine · Circuit

W. Zhang (✉)
National Institute on Drug Dependence, Peking University, Beijing, China
e-mail: zhangw@hsc.pku.edu.cn

4.1 Introduction

Innate behaviors, programmed by genetically determined neural circuits, are stereotyped and robust to be observed in generations of the same species. While it is believed to be "hardwired," which indicates it is hard to change, however, those behaviors are often found to be flexible and subjected to both internal and external environment states or contexts. For example, most animals exhibit ingestional neophobia when exposed to a new food, that is animals would only eat a little of this novel food presented to them (Domjan et al. 1977). This behavior has significant advantage to those animals as many foods in natural environment is poisonous or indigestible. By limited consumption of food, animals could value whether food can be safely consumed and not detrimental to their health. After multiple exposure and consumption of the food, the ingestional neophobia could subside and the animal consumes greater amount of food. This behavior change to overcome ingestional neophobia is advantageous for animal since it promotes energy storage and prevents energy deficit or hunger. Besides it satisfies basic needs of the animal, food also is a potent reward to promote change in the feeding behavior.

Studies have identified neurons in multiple brain structures that have a critical role in reward processing and behavior change. This brain regions also form connections between them and form an important brain circuit often called

H. Wang (ed.), *Neural Circuits of Innate Behaviors*, Advances in Experimental Medicine and Biology 1284, https://doi.org/10.1007/978-981-15-7086-5_4

"reward pathway" (Koob and Volkow 2016). Reward pathway includes ventral tegmental area (VTA), substantia nigra, striatum, prefrontal cortex, orbitofrontal cortex, amygdala, hippocampus, and associated structures (Fig. 4.1). Among these brain regions, VTA is important for reward and motivation, ventral striatum is important for habitation and locomotion, amygdala has a critical role in fear and negative emotions, and the cortical regions are important for processing memories and emotions. One of the best characterized circuit in the reward pathway is dopaminergic projection from the ventral tegmental area (VTA) to the nucleus accumbens (NAc), which localized at the ventral part of the striatum. In NAc, the major type of neurons is medium spiny neurons (MSNs), which is GABAergic and comprises more than 90% of neurons in NAc (Kreitzer 2009). MSNs can be further categorized into two groups: those that expressing dopamine receptor subtype 1 (D1-MSNs) and those that expressing dopamine receptor subtype 2 (D2-MSNs). Those two types of MSNs not only express different dopamine receptor subtypes, the intracellular downstream molecules activated by dopamine receptor activation are also different, and furthermore they project to different brain regions. For example, D1-MSNs in NAc directly innervate the VTA (Xia et al. 2011), whereas D2-MSNs innervate the VTA by impeding GABAergic neurons in the ventral pallidum (Soares-Cunha et al. 2018). Beside projections to the NAc, VTA dopaminergic neurons also project to several other brain regions in the reward pathway, such as the prefrontal cortex, central and basolateral amygdala, and the hippocampus (Robison and Nestler 2011). These brain regions in reward pathway are also interconnected besides dopaminergic innervation from VTA. For example, NAc receives extensive glutamatergic inputs from the PFC, ventral hippocampus, and amygdala. And the PFC, hippocampus, and amygdala form reciprocal excitatory glutamatergic projections with one another. Besides the three neurotransmitters mentioned above, brain regions in the reward pathway are also modulated by cholinergic interneurons in those regions, the serotonergic inputs from dorsal raphe, and noradrenergic inputs from locus coeruleus. Furthermore, studies have shown that the VTA dopaminergic neurons could also co-release glutamate or GABA with dopamine at terminals (Tritsch et al. 2012; Hnasko et al. 2012).

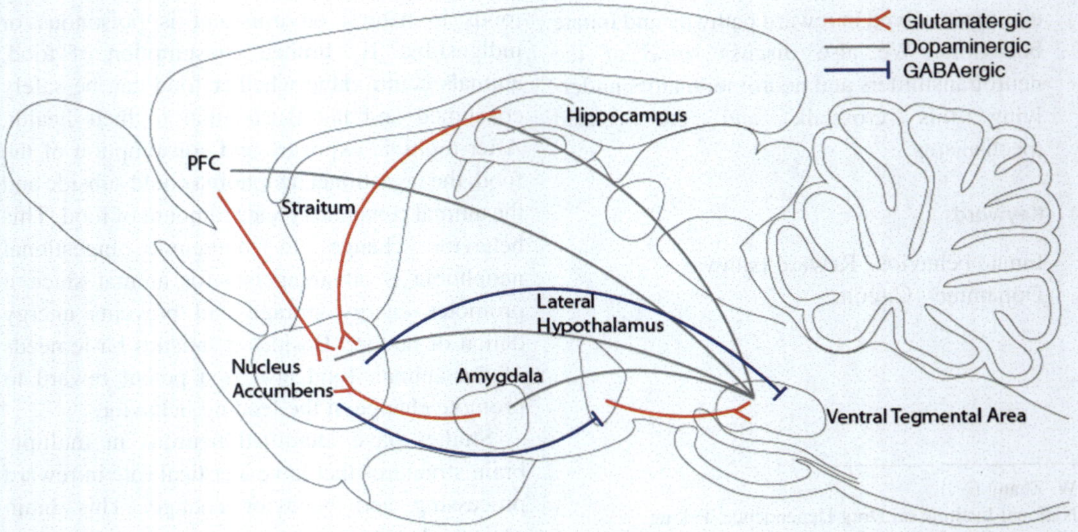

Fig. 4.1 The reward pathway. A simplified diagram of the reward pathway and brain regions involved in the rodent brain. The major glutamatergic, dopaminergic, and GABAergic projections between those brain regions are illustrated

4.2 Dopamine and Innate Behavior

In reward pathway, dopamine is a key neurotransmitter to modulate reward. The aforementioned VTA–NAc circuit has a crucial role in the reward recognition and subsequently initiating reward consumption (Koob and Le Moal 2008), and this circuit is all important in brain response to aversive stimuli as well. In the brain, dopamine is classically viewed as an enforcer for motivation and reinforcement (Wise 2004). Dopamine was first identified with motivational function by a study showing that damage to the nigrostriatal dopamine fibers led to feeding and drinking deficits (Robbins et al. 1986), and damage to the mesolimbic dopamine fibers decreased locomotion (Cools 1986). Dopamine is also shown to be the key neurotransmitter in the reinforcement learning (Schultz 1998). Animals do not learn to lever-press for food or water if dopamine function is impaired during training, and blockage of dopamine system led to decline in performance in well-trained animals (Wise and Schwartz 1981). By this role of dopamine, it could modulate animal behaviors, even those "hardwired" innate behaviors. Indeed, under some pathological conditions, change of dopamine in the brain led to suppression of innate behaviors. For example, cocaine addiction is one of the most prevalent addictions globally. Cocaine inhibits dopamine uptake, leads to buildup of extracellular dopamine (Giros et al. 1996), which is viewed as the basis for addiction. Cocaine addiction not only led to change in abnormal goal-directed learning and habit formation but also suppressed neophobia in rats to novel objects in the environment (Stansfield and Kirstein 2007) and novel food (Goudie et al. 1978).

A study on human subject showed increased dopamine release in dorsal striatum when palatable food is ingested is correlated with self-reported pleasure from eating the food (Small et al. 2003). These studies indicate that reward pathway, especially dopamine release in brain regions of the reward pathway, could modulate innate behaviors. Indeed, in rodents lesion of the dopamine innervation in medial shell of the nucleus accumbens, which is part of the ventral striatum, led to elongated neophobia to sucrose (Martinez-Hernandez et al. 2012). Meanwhile, the same female mice showed no change in the preference to sexual hormone of male mice after lesion (Martinez-Hernandez et al. 2012). These results suggest that reward pathway might only affect some kinds of innate behaviors and the activity of their related circuits. Besides foods, animals often show neophobia to other novel stimuli, such as novel objects introduced into their environment. Studies have shown a similar important role of striatum and dopamine in modulation of this behavior. Lesion of dorsal striatum also reduced neophobia to novel object (Cigrang et al. 1986). Knockdown of dopamine transporter (*Dat1*) gene in mice increased extracellular dopamine in the brain, and mice showed enhanced investigation of objects (Pogorelov et al. 2005).

Although studies have shown the possible roles of dopamine and reward pathway in innate behaviors, however, the related field only recently begin to dissect the circuit in reward pathway involved in innate behavior. Recently this question begins to be addressed by researchers utilizing genetic tools, especially optogenetic method. In below, we summarized recent progress on this subject.

4.3 Medium Spiny Neurons in Striatum and Innate Behavior

In striatum, the majority of neurons are GABAergic medium spiny neurons (MSNs). More than 90% neurons in striatum are MSNs. Striatal MSNs can be classified into two groups, those expressing dopamine receptor D1-like and others expressing dopamine receptor D2-like. Dopamine receptors are G-protein-coupled receptors, D1-like dopamine receptor includes D1 and D5 receptor subtypes, D2-like dopamine receptor includes D2, D3, and D4 subtypes (Beaulieu and Gainetdinov 2011). Those two classes have opposite influence on adenylyl cyclase, and the MSNs expressing those two classes also have opposite functions on animal behavior. Generally, MSNs expressing D1-like receptors (D1-MSNs) promote action, while those expressing D2-like receptors (D2-MSNs) inhibit action. Besides dopamine receptor subtype difference, those two types of MSNs are involved in different circuits. D1-MSNs directly project to GPi/SNc, while D2-MSNs project to GPI/SNc relays in the globus pallidus and subthalamus. Thus D1-MSNs are involved in "direct pathway," and D2-MSNs are in "indirect pathway." The output of striatum is generally viewed as to reflect the balance between these two projections (Kreitzer 2009). Between these two types of MSNs, activation of D2-MSNs is shown to be vital in innate risk-avoidance task. Inhibition of D2-MSNs via Chemogenetic modulation reduced innate risk avoidance of mice to the odor of fox urine (Blomeley et al. 2018). Those D2-MSNs also are modulated by orexins (hypocretin) released from hypothalamic neurons (Gutierrez et al. 2011). Orexin excited D2-MSNs in ventral striatum, but not D1-MSNs or interneurons (Blomeley et al. 2018). Orexin is required for context-dependent brain control of behavior, and orexin-releasing neurons are activated by internal and external stresses, and orexin release is necessary for inducing anxiety-like states, such as increased innate risk avoidance in animals or panic attack in human subjects (Johnson et al. 2010; Suzuki et al. 2005). Application of orexin receptor antagonist reduced risk avoidance in mice. The finding that orexin-releasing targets D2-MSNs in the ventral striatum, suggests a possible mechanism of crosstalk between circuits controlling innate behavior and reward pathway. However, there still is discrepancy between studies as one reported no effect of orexin-releasing output from hypothalamic neurons on D1-MSNs, while others reported orexin excited more than 80% of ventral striatum neurons (Mori et al. 2011). More than 90% of ventral striatum neurons are MSNs, and the two types of MSNs, D1-MSNs and D2-MSNs, roughly each represents half of the MSNs in striatum. This implies more delicate dissection of this circuit is necessary to understand this circuit and its function.

4.4 Ventral Tegmental Area (VTA) Neurons and Innate Behavior

Among brain regions involved in reward pathway, VTA is one of the most important structure which includes dopaminergic neurons that project to many other brain regions in the reward pathway. Studies have shown that VTA not only contains projecting dopaminergic neurons but also GABAergic and glutamatergic neurons (Yamaguchi et al. 2007; Dobi et al. 2010). Those GABAergic neurons not only inhibit dopaminergic neurons in VTA, but they also project to other brain regions (Beier et al. 2015). A recent study showed that besides projections to nucleus accumbens (Brown et al. 2012), GABAergic neurons in VTA also send long projection to the central nucleus of the amygdala (CeA), lateral hypothalamus, lateral habenula, and the periaqueductal gray (Zhou et al. 2019). CeA is important for the innate defensive behaviors of rodents (Isosaka et al. 2015), and the VTA GABAergic neurons majorly project to the medial CeA (Zhou et al. 2019). In the presence of innate looming threat that mimic the diving movement of mouse predators, such as hawks, the VTA GABAergic projection neurons activated and inhibited CeA neurons which promoted defensive behavior of mice as blocking GABAergic transmission in CeA delayed the latency of looming-evoked defensive-like behavior. This circuit is enervated by projections from superior colliculus, which conveys visual information to VTA neurons (Zhou et al. 2019).

4.5 Prefrontal Cortex and Innate Behavior

Among brain regions in reward pathway, the prefrontal cortex (PFC) is important for multiple vital brain functions, such as decision-making, attention, and working memory. Recent studies also showed modification of innate behavior by prefrontal activity change. In cortex, besides parvalbumin- and somatostatin-expressing neurons (Zhang et al. 2016; Zhang et al. 2017;

Kawaguchi 1993), neuropeptide Y-expressing neurons are also GABAergic neurons; they made up less than 10% of GABAergic interneuron in cortex, and they also often express somatostatin (Kubota et al. 2011). Reduction of GABA synthesis in neuropeptide Y-expressing (NPY+) GABAergic neurons of PFC enhanced innate behaviors, such as anxiety-like activity, nesting construction, and social dominance, but not innate fear expression (Corder et al. 2018). Besides GABAergic inhibition, changes of D2-like dopamine receptor subtype 4 (D4R) in PFC also could modulate innate fear expression. Inhibition of D4R reduced fear expression when rats were exposed to cat odor (Vergara et al. 2017). D4R is expressed in GABAergic neurons, including parvalbumin-positive (PV+) interneuron, suggesting the involvement of multiple types of GABAergic neurons in the modulation of innate behaviors by PFC. Interestingly, these studies suggest that NPY+ and PV+ interneurons might be involved in different innate behaviors, whether these results suggest different circuits that these two types of neurons are involved in is not clear.

Anterior cingulate cortex (ACC) locates rostral in the prefrontal lobe area. It is well-known for its role in emotion, pain, and cognitive control, and studies have also shown that it plays a critical role in regulating fear responses (Shackman et al. 2011). The ACC is extensively interconnected with limbic nuclei including the amygdala, hippocampus, and ventral striatum (Cassell and Wright 1986; Christie et al. 1987; Reep and Corwin 1999). In amygdala, glutamatergic projections from ACC innervate both basolateral (BLA) and central nucleus (CeA) of amygdala, while the glutamatergic input from ACC to CeA is very sparse (Jhang et al. 2018; McDonald et al. 1996). As mentioned before, CeA is important for the innate defensive behaviors, while BLA is shown to be involved in innate freezing behavior (Vazdarjanova et al. 2001). Modulation of the ACC-BLA circuit function contributes to innate fear response to a predator odor in mice. Inactivation of this projection enhanced freezing response to fox urine without affecting conditioned freezing, meanwhile excitation of ACC projection robustly inhibits both innate and conditioned freezing (Jhang et al. 2018).

4.6 Conclusion

It is well established that dopamine and reward pathway in the brain are important in goal-directed behaviors. Besides this, this complexed neurotransmitter and circuit system is also shown to be the circuits and homeostatic signals that control hunger, satiety, and motivations. Sitting at the crossroad for both environment and innate behavior, we would expect more researches to address the important role of this system in the crosstalk of nature with nurture. Furthermore, our understanding of the role of the dopamine system and reward pathway in innate behaviors will not only help us understand the environment role in behavior but also shed light on novel interventions to symptoms of psychologic and neurological diseases.

Acknowledgment This work was supported by National Natural Science Foundation of China Grants 31741060 and 91732109.

References

Beaulieu JM, Gainetdinov RR (2011) The physiology, signaling, and pharmacology of dopamine receptors. Pharmacol Rev 63(1):182–217

Beier KT et al (2015) Circuit architecture of VTA dopamine neurons revealed by systematic input-output mapping. Cell 162(3):622–634

Blomeley C, Garau C, Burdakov D (2018) Accumbal D2 cells orchestrate innate risk-avoidance according to orexin signals. Nat Neurosci 21(1):29–32

Brown MT et al (2012) Ventral tegmental area GABA projections pause accumbal cholinergic interneurons to enhance associative learning. Nature 492 (7429):452–456

Cassell MD, Wright DJ (1986) Topography of projections from the medial prefrontal cortex to the amygdala in the rat. Brain Res Bull 17(3):321–333

Christie MJ et al (1987) Excitatory amino acid projections to the nucleus accumbens septi in the rat: a retrograde transport study utilizing D[3H]aspartate and [3H] GABA. Neuroscience 22(2):425–439

Cigrang M, Vogel E, Misslin R (1986) Reduction of neophobia in mice following lesions of the caudate-putamen. Physiol Behav 36(1):25–28

Cools AR (1986) Mesolimbic dopamine and its control of locomotor activity in rats: differences in pharmacology and light/dark periodicity between the olfactory tubercle and the nucleus accumbens. Psychopharmacology 88(4):451–459

Corder KM et al (2018) Prefrontal cortex-dependent innate behaviors are altered by selective knockdown of Gad1 in neuropeptide Y interneurons. PLoS One 13(7): e0200809

Dobi A et al (2010) Glutamatergic and nonglutamatergic neurons of the ventral tegmental area establish local synaptic contacts with dopaminergic and nondopaminergic neurons. J Neurosci 30(1):218–229

Domjan M, Schorr R, Best M (1977) Early environmental influences on conditioned and unconditioned ingestional and locomotor behaviors. Dev Psychobiol 10(6):499–506

Giros B et al (1996) Hyperlocomotion and indifference to cocaine and amphetamine in mice lacking the dopamine transporter. Nature 379(6566):606–612

Goudie AJ, Dickins DW, Thornton EW (1978) Cocaine-induced conditioned taste aversions in rats. Pharmacol Biochem Behav 8(6):757–761

Gutierrez R et al (2011) Neural integration of reward, arousal, and feeding: recruitment of VTA, lateral hypothalamus, and ventral striatal neurons. IUBMB Life 63 (10):824–830

Hnasko TS et al (2012) Ventral tegmental area glutamate neurons: electrophysiological properties and projections. J Neurosci 32(43):15076–15085

Isosaka T et al (2015) Htr2a-expressing cells in the central amygdala control the hierarchy between innate and learned fear. Cell 163(5):1153–1164

Jhang J et al (2018) Anterior cingulate cortex and its input to the basolateral amygdala control innate fear response. Nat Commun 9(1):2744

Johnson PL et al (2010) A key role for orexin in panic anxiety. Nat Med 16(1):111–115

Kawaguchi Y (1993) Groupings of nonpyramidal and pyramidal cells with specific physiological and morphological characteristics in rat frontal cortex. J Neurophysiol 69(2):416–431

Koob GF, Le Moal M (2008) Addiction and the brain antireward system. Annu Rev Psychol 59:29–53

Koob GF, Volkow ND (2016) Neurobiology of addiction: a neurocircuitry analysis. Lancet Psychiatry 3 (8):760–773

Kreitzer AC (2009) Physiology and pharmacology of striatal neurons. Annu Rev Neurosci 32:127–147

Kubota Y et al (2011) Selective coexpression of multiple chemical markers defines discrete populations of neocortical GABAergic neurons. Cereb Cortex 21 (8):1803–1817

Martinez-Hernandez J, Lanuza E, Martinez-Garcia F (2012) Lesions of the dopaminergic innervation of the nucleus accumbens medial shell delay the generation of preference for sucrose, but not of sexual pheromones. Behav Brain Res 226(2):538–547

McDonald AJ, Mascagni F, Guo L (1996) Projections of the medial and lateral prefrontal cortices to the amygdala: a Phaseolus vulgaris leucoagglutinin study in the rat. Neuroscience 71(1):55–75

Mori K, Kim J, Sasaki K (2011) Electrophysiological effects of orexin-B and dopamine on rat nucleus accumbens shell neurons in vitro. Peptides 32 (2):246–252

Pogorelov VM et al (2005) Novelty seeking and stereotypic activation of behavior in mice with disruption of the Dat1 gene. Neuropsychopharmacology 30 (10):1818–1831

Reep RL, Corwin JV (1999) Topographic organization of the striatal and thalamic connections of rat medial agranular cortex. Brain Res 841(1–2):43–52

Robbins JA, Logemann JA, Kirshner HS (1986) Swallowing and speech production in Parkinson's disease. Ann Neurol 19(3):283–287

Robison AJ, Nestler EJ (2011) Transcriptional and epigenetic mechanisms of addiction. Nat Rev Neurosci 12 (11):623–637

Schultz W (1998) Predictive reward signal of dopamine neurons. J Neurophysiol 80(1):1–27

Shackman AJ et al (2011) The integration of negative affect, pain and cognitive control in the cingulate cortex. Nat Rev Neurosci 12(3):154–167

Small DM, Jones-Gotman M, Dagher A (2003) Feeding-induced dopamine release in dorsal striatum correlates with meal pleasantness ratings in healthy human volunteers. NeuroImage 19(4):1709–1715

Soares-Cunha C et al (2018) Nucleus accumbens microcircuit underlying D2-MSN-driven increase in motivation. eNeuro 5(2):ENEURO.0386-18.2018

Stansfield KH, Kirstein CL (2007) Chronic cocaine or ethanol exposure during adolescence alters novelty-related behaviors in adulthood. Pharmacol Biochem Behav 86(4):637–642

Suzuki M et al (2005) Orexin-A (hypocretin-1) is possibly involved in generation of anxiety-like behavior. Brain Res 1044(1):116–121

Tritsch NX, Ding JB, Sabatini BL (2012) Dopaminergic neurons inhibit striatal output through non-canonical release of GABA. Nature 490(7419):262–266

Vazdarjanova A, Cahill L, McGaugh JL (2001) Disrupting basolateral amygdala function impairs unconditioned freezing and avoidance in rats. Eur J Neurosci 14 (4):709–718

Vergara MD et al (2017) Activation of type 4 dopaminergic receptors in the prelimbic area of medial prefrontal cortex is necessary for the expression of innate fear behavior. Behav Brain Res 324:130–137

Wise RA (2004) Dopamine, learning and motivation. Nat Rev Neurosci 5(6):483–494

Wise RA, Schwartz HV (1981) Pimozide attenuates acquisition of lever-pressing for food in rats. Pharmacol Biochem Behav 15(4):655–656

Xia Y et al (2011) Nucleus accumbens medium spiny neurons target non-dopaminergic neurons in the ventral tegmental area. J Neurosci 31(21):7811–7816

Yamaguchi T, Sheen W, Morales M (2007) Glutamatergic neurons are present in the rat ventral tegmental area. Eur J Neurosci 25(1):106–118

Zhang W et al (2016) Hyperactive somatostatin interneurons contribute to excitotoxicity in neurodegenerative disorders. Nat Neurosci 19(4):557–559

Zhang W et al (2017) BDNF rescues prefrontal dysfunction elicited by pyramidal neuron-specific DTNBP1 deletion in vivo. J Mol Cell Biol 9(2):117–131

Zhou Z et al (2019) A VTA GABAergic neural circuit mediates visually evoked innate defensive responses. Neuron 103:473

Neuronal Response and Behavioral Modulation in Social Interactions

5

Yang Zhan

Abstract

Social behavior is a complex behavior that requires processing of sensory cues and integration of internal states. Social interaction involves two or more individuals to approach each other and engage communications. Although sensory, motivational, emotional, or reward cues may all play roles in directing the sociability and social preference during social interaction, how neural activities from different brain regions are modulated during the behavioral process of social interaction are only beginning to be studied. Multiple brain regions including prefrontal cortex, hippocampus, and amygdala contain active neurons during social interaction. This review examines the neural responses in behaving rodents during social behavior and discusses how manipulation of specific neural pathways can modulate social behavior. Neural activities during social interaction provide direct measurements about how social information is coded and are beneficial in understanding the neural mechanisms underlying social behavior.

Keywords

Social interaction · Prefrontal cortex · Hippocampus · Neural coding · In vivo neural recordings

Social interaction refers to a process of reciprocal stimulation or response between two or multiple individuals. In this review, I focus on the brain areas that are responsive during the social interaction. Emphasis is put on the single-unit activity of the neurons so that evidence of the involvement of the neural circuits about encoding of socially relevant information is given. Next, I put together the recent results on behavioral manipulation of social behavior using circuitry manipulation tools, mainly based on optogenetic or pharmacogenetic approaches. Much of the knowledge about how the brain substrates underly the social behavior come from mouse models. Hopefully the fundamental studies on social interaction will contribute to the understanding of behavioral deficits associated with abnormal social behavior.

Y. Zhan (✉)
Brain Cognition and Brain Disease Institute, Shenzhen Institutes of Advanced Technology, Chinese Academy of Sciences, Shenzhen, Guangdong, China
e-mail: yang.zhan@siat.ac.cn

H. Wang (ed.), *Neural Circuits of Innate Behaviors*, Advances in Experimental Medicine and Biology 1284, https://doi.org/10.1007/978-981-15-7086-5_5

5.1 Neuronal Response in Social Behavior

5.1.1 Neuronal Response in Social Interaction

5.1.1.1 Medial Prefrontal Cortex

In the social interaction, medial prefrontal cortex (mPFC) plays important roles (Bicks et al. 2015). In the mouse behavioral tests, social interaction can be measured when the test mouse approaches a stimulus mouse and they engage interaction. The test mouse takes the initiative to approach the target stimulus, and it touches the other mouse on the face or other body parts by the nose. The mPFC single units recorded by the extracellular electrophysiology showed response during the nose poke (Lee et al. 2016). Analysis of the social approach period shows that neural activities show difference between investigation of the social stimulus and the object stimulus. The proportion of the neurons that were found to be responsive is about 15%. Using miniature fluorescence microscope and GCamp indicators, mPFC neurons showed response during two consecutive stages of sociality and social novelty (Liang et al. 2018). Both increased and decreased neurons were found. The specific responses to the social stimuli demonstrate that mPFC neurons can code information during approach to social target. It seems that mPFC neurons show overlapping responses to familiar and novel social stimuli, probably due to the properties of mixed selectivity (Rigotti et al. 2013; Parthasarathy et al. 2017).

PFC has many outputs to the other brain regions (Hoover and Vertes 2007). With virus labeling tools, the PFC neurons projecting to the specific downstream targets have been investigated. In three-chamber tasks, it was found that PFC neurons projecting to the nucleus accumbens displayed social response or spatial response. This probably indicates that PFC-nucleus accumbens pathway contains either social or spatial information, or a combination of both (Murugan et al. 2017).

5.1.1.2 Amygdala

Amygdala is an area that is responsive to socially relevant stimuli (Adolphs 2009). A recent study described the neuronal activities in the medial amygdala (MeA) during social interaction using miniature fluorescence microscope (Li et al. 2017). A sizable proportion of neurons showed increased and decreased response to the social stimuli. In both male and female mice, the proportion of the responsive neurons is similar. Interestingly, after the mice had sexual experience, the number of neurons responsive to the social stimuli became higher.

5.1.1.3 Ventral Tegmental Area

Ventral tegmental area (VTA) area is involved in processing emotionally salient stimuli. Using fiber photometry approach measuring the bulk calcium signals in the VTA, it has been found that VTA dopamine positive neurons responded to social stimuli when another social stimulus was introduced to a homecage with test mouse (Gunaydin et al. 2014). As the test mice repeatedly investigated the social targets, the calcium signals became less prominent.

5.1.2 Neuronal Response in Aggression

5.1.2.1 Hypothalamus

In aggressive behavior, mice can attack other conspecifics or objects by biting. The aggressive behavior is instinctive and threatening cues from external stimuli can elicit aggression. The electrophysiological single-unit recordings from the hypothalamus have found neurons that were responsive during male–male attack (Lin et al. 2011; Falkner et al. 2014). The ventrolateral subdivision of ventromedial hypothalamus (VMH) contains neurons that respond exclusively during attack. The male responsive neurons seemed to be selective since many of them remained silent to the female stimulus.

5.1.3 Neuronal Response in Dominance

5.1.3.1 mPFC

Dominance behavior is reflected by hierarchical structure in a group of mice. The tube test can measure the dominance hierarchy in which a dominant mouse can push out the subordinate one in a tube contest. mPFC has been found to be important in dominance behavior measured in the tube test (Wang et al. 2011; Zhou et al. 2017). In the tube test, mPFC neurons in the dominant mouse display elevated response related to the push and effortful behavior during the contest. Mice with established dominance hierarchy may utilize the critical information processing within the mPFC to win the contest. Furthermore, it has been shown that the mPFC plasticity depends on the inputs from the mediodorsal thalamus.

5.1.4 Neuronal Response in Social Defeat

In contrast to the dominant mouse, the subordinate mouse can display behavioral and neural changes. mPFC is thought to play roles in social defeat behavior. Using ΔFosB labeling technique which indicates neuronal activity changes under chronic conditions, mPFC was found to be involved after social defeat (Hinwood et al. 2010). With local field potential recordings, mPFC and dorsal periaqueductal gray (dPAG) synchronization has been examined after social defeat (Franklin et al. 2017). A subchronic 3-day social defeat caused a reduction in the coherence between the mPFC and the dPAG, demonstrating that the functional connectivity of this pathway was compromised. Further examination of the directed relationship by Granger causality (Zhan 2015) between the dPAG and mPFC reflected that social defeat resulted in an increased information flow from the dPAG to the mPFC. The descending mPFC to dPAG pathway therefore can process the stress-related information induced by the social defeat.

5.1.5 Neuronal Response in Social Memory

Social memory reflects the ability of the mice to recognize a new conspecific from a remembered one. The social memory test comprises the comparison of interacting time between a previously encountered stimulus mouse and a novel stimulus mouse. During the preference test in which both the familiar mouse and the novel mouse are present, the test mouse is considered to form social memory if more visits are spent with the novel mouse. Using a long-term social memory test and *c-fos* approach (Lüscher Dias et al. 2016), it was found that amygdala, prelimbic area of the PFC, and hippocampus can contribute to the representation of social memory.

5.1.5.1 Hippocampus

Ventrial hippocampus (vHPC) is considered to process different information from the dHPC. vHPC is involved in anxiety and emotion. In social memory test, vHPC has been found to be activated by a remembered mouse (Okuyama et al. 2016). By miniature fluorescence microscope, it was found that vHPC had a higher proportion of neurons responding to a social stimulus after familiarization with it. In contrast, neurons in the dorsal hippocampus (dHPC) did not show increased proportion. vHPC can hold the information of a familiar social stimulus.

During facial whisker contact, vHPC neurons from the rats display elevated response (Rao et al. 2019) when the stimulus rat was in presence compared to the periods when the test rat was alone. The response in the vHPC were similar when the same conspecific was subject to repeated presentations. In male rats, female conspecifics evoked stronger responses than males. The dHPC was not found to be socially responsive (von Heimendahl et al. 2012). Therefore, vHPC contains socially relevant information.

Hippocampal CA2 region is involved in social processing. Genetic lesion of this region has been implicated in social memory (Hitti and Siegelbaum 2014). Recordings of the rat CA2 neurons did not reveal a marked change on the firing rates when the test mouse was exposed with social stimuli (Alexander et al. 2016). Although there was lack of social response for CA2 neurons, spatial correlations before and after the social stimulus were different. The social stimulus can modify the spatial representation in the hippocampus.

Parvalbumin positive neurons in the vHPC were found to be responsive to the social stimuli (Deng et al. 2019). Using fiber photometry and transgenic mice labeling the parvalbumin neurons, the bulk calcium signals of the labeled neurons had a stronger response to the novel stimuli, compared to the familiar stimuli. The distinct response of the parvalbumin neurons in the vHPC demonstrates that this class of neurons may play important roles for distinguishing novel mice and familiar mice.

5.2 Neural Circuit Manipulation and Social Behavior

5.2.1 Social Interaction

5.2.1.1 mPFC

mPFC has extensive input and output brain structures. Prefrontal lesions of the nicotinic receptors produced abnormal social interaction behavior with increased investigation time (Avale et al. 2011). Conditional knockout of *SHANK3* gene in the anterior cingulate cortex produced social interaction deficits (Guo et al. 2019). The requirements of mPFC neural functions contribute to the normal expression of social interaction in mice. The mPFC-nucleus accumbens projections were found to decrease the social interaction time (Murugan et al. 2017).

5.2.1.2 VTA

VTA and the VTA to nucleus accumbens pathway have been investigated in social interaction (Gunaydin et al. 2014). Social interaction may involve processing of stimuli with positive valence. Optogenetic stimulation of channelrhodopsin-2 (ChR2) expressed in tyrosine hydroxylase (TH)-positive neurons in the VTA promoted social interaction in homecage direct social interaction assay (Gunaydin et al. 2014). When these neurons were inhibited with halorhodopsin (eNpHR3.0), the social interaction time was reduced. Therefore, it seems that VTA dopamine neurons have a modulatory role for social behavior. Furthermore, activation of VTA to NAc pathway increased social interaction in the three-chamber social interaction assay.

5.2.1.3 Amygdala

Basolateral complex of amygdala (BLA) to vHPC pathway has been investigated in social interaction (Felix-Ortiz and Tye 2014). Using optogenetic method, it has been shown that inhibiting BLA-vHPC pathway increased social interaction time in a resident-intruder assay. On the other hand, activation of the BLA-vHPC pathway decreased the social interaction. Although both structures of the BLA and vHPC are involved in the social behavior, how the interaction in the BLA-vHPC connectivity modulates social behavior needs further investigation.

5.2.2 Social Memory

5.2.2.1 Hippocampus

Manipulation of the CA2 and vHPC has demonstrated that these regions can modulate social interaction. Genetic targeting of the pyramidal cells in the dorsal CA2 has shown that social memory was impaired when these cells were ablated (Hitti and Siegelbaum 2014) though sociability or other spatial memory was not affected. Furthermore, it has been shown that dorsal CA2 projections to the ventral CA1 can modulate the encoding, reconsolidation, and retrieval during the social memory processes (Meira et al. 2018). Using *c-fos* and optogenetic labeling techniques, it has been demonstrated that reactivation of the labeled ventral CA1 neurons contributes to the memory retrieval of the previously encounter mouse (Okuyama et al. 2016). In the ventral CA1, social isolation caused reduction of parvalbumin neurons and ablating parvalbumin neurons resulted in impairments of the social memory (Deng et al. 2019). These studies about the functions of CA2 and the vHPC in social memory have underlined the role of the hippocampus in social memory.

5.2.3 Dominance Behavior

5.2.3.1 mPFC

Direct mPFC synaptic manipulation has shown that decreasing the synaptic efficacy within the mPFC can lead to the hierarchy changes in a group of mice in the tube test (Wang et al. 2011). Further, thalamic inputs from the medial dorsal thalamus (MDT) to the mPFC can modulate the hierarchy changes (Zhou et al. 2017). This is dependent on the long-term potentiation of this MDT-mPFC pathway. Optogenetic manipulation of the mPFC during the social contest in the tube test has shown that activation of the pyramidal neurons in the mPFC can induce instant wining thus augmenting the social hierarchy.

5.2.4 Aggression

5.2.4.1 Hypothalamus

The ventrolateral part of the ventromedial hypothalamus (VMHvl) is involved in mediating aggressive behavior. Optogenetic activation of VMHvl induces attack behavior and inactivation of this region suppressed aggression (Falkner et al. 2016). In female mice, it has also been shown that activation of VMHvl caused attacks toward other targets (Hashikawa et al. 2017). Therefore, VMHvl may play a prominent role in the regulation of aggressive behavior.

To summarize, social interaction forms fundamental basis for establishing social relationships in species living in a group. It contributes to recognition, cooperation, and competition. Social interaction can lead to positive social and emotional development. On the other hand, deficient or lack of social interaction can have negative impacts on physiological or neurological development. The study of the brain substrates underlying the social interaction can help to understand how this important behavior is modulated. Although social interaction integrates complex cues of sensory information or internal processing, new tools and novel findings about this behavior is beginning to bring in new advances in this area.

References

Adolphs R (2009) The social brain: neural basis of social knowledge. Annu Rev Psychol 60:693–716

Alexander GM, Farris S, Pirone JR, Zheng C, Colgin LL, Dudek SM (2016) Social and novel contexts modify hippocampal CA2 representations of space. Nat Commun 7:10300

Avale ME, Chabout J, Pons S, Serreau P, De Chaumont F, Olivo-Marin J-C, Bourgeois J-P, Maskos U, Changeux J-P, Granon S (2011) Prefrontal nicotinic receptors control novel social interaction between mice. FASEB J 25:2145–2155

Bicks LK, Koike H, Akbarian S, Morishita H (2015) Prefrontal cortex and social cognition in mouse and man. Front Psychol 6:1805

Deng X, Gu L, Sui N, Guo J, Liang J (2019) Parvalbumin interneuron in the ventral hippocampus functions as a discriminator in social memory. Proc Natl Acad Sci 116:16583–16592

Falkner AL, Dollar P, Perona P, Anderson DJ, Lin D (2014) Decoding ventromedial hypothalamic neural activity during male mouse aggression. J Neurosci 34:5971

Falkner AL, Grosenick L, Davidson TJ, Deisseroth K, Lin D (2016) Hypothalamic control of male aggression-seeking behavior. Nat Neurosci 19:596–604

Felix-Ortiz AC, Tye KM (2014) Amygdala inputs to the ventral hippocampus bidirectionally modulate social behavior. J Neurosci 34:586–595

Franklin TB, Silva BA, Perova Z, Marrone L, Masferrer ME, Zhan Y, Kaplan A, Greetham L, Verrechia V, Halman A, Pagella S, Vyssotski AL, Illarionova A, Grinevich V, Branco T, Gross CT (2017) Prefrontal cortical control of a brainstem social behavior circuit. Nat Neurosci 20:260–270

Gunaydin LA, Grosenick L, Finkelstein JC, Kauvar IV, Fenno LE, Adhikari A, Lammel S, Mirzabekov JJ, Airan RD, Zalocusky KA, Tye KM, Anikeeva P, Malenka RC, Deisseroth K (2014) Natural neural projection dynamics underlying social behavior. Cell 157:1535–1551

Guo B, Chen J, Chen Q, Ren K, Feng D, Mao H, Yao H, Yang J, Liu H, Liu Y, Jia F, Qi C, Lynn-Jones T, Hu H, Fu Z, Feng G, Wang W, Wu S (2019) Anterior

cingulate cortex dysfunction underlies social deficits in Shank3 mutant mice. Nat Neurosci 22:1223–1234
Hashikawa K, Hashikawa Y, Tremblay R, Zhang J, Feng JE, Sabol A, Piper WT, Lee H, Rudy B, Lin D (2017) Esr1+ cells in the ventromedial hypothalamus control female aggression. Nat Neurosci 20:1580–1590
von Heimendahl M, Rao RP, Brecht M (2012) Weak and nondiscriminative responses to conspecifics in the rat Hippocampus. J Neurosci 32:2129–2141
Hinwood M, Tynan RJ, Day TA, Walker FR (2010) Repeated social defeat selectively increases ΔFosB expression and histone H3 acetylation in the infralimbic medial prefrontal cortex. Cereb Cortex 21:262–271
Hitti FL, Siegelbaum SA (2014) The hippocampal CA2 region is essential for social memory. Nature 508:88–92
Hoover W, Vertes R (2007) Anatomical analysis of afferent projections to the medial prefrontal cortex in the rat. Brain Struct Funct 212:149–179
Lee E, Rhim I, Lee JW, Ghim J-W, Lee S, Kim E, Jung MW (2016) Enhanced neuronal activity in the medial prefrontal cortex during social approach behavior. J Neurosci 36:6926–6936
Li Y, Mathis A, Grewe BF, Osterhout JA, Ahanonu B, Schnitzer MJ, Murthy VN, Dulac C (2017) Neuronal representation of social information in the medial amygdala of awake behaving mice. Cell 171:1176–1190.e1117
Liang B, Zhang L, Barbera G, Fang W, Zhang J, Chen X, Chen R, Li Y, Lin D-T (2018) Distinct and dynamic ON and OFF neural ensembles in the prefrontal cortex code social exploration. Neuron 100:700–714.e709
Lin D, Boyle MP, Dollar P, Lee H, Lein ES, Perona P, Anderson DJ (2011) Functional identification of an aggression locus in the mouse hypothalamus. Nature 470:221–226
Lüscher Dias T, Fernandes Golino H, Oliveira VEMD, Dutra Moraes MF, Schenatto Pereira G (2016) C-Fos expression predicts long-term social memory retrieval in mice. Behav Brain Res 313:260–271
Meira T, Leroy F, Buss EW, Oliva A, Park J, Siegelbaum SA (2018) A hippocampal circuit linking dorsal CA2 to ventral CA1 critical for social memory dynamics. Nat Commun 9:4163
Murugan M, Jang HJ, Park M, Miller EM, Cox J, Taliaferro JP, Parker NF, Bhave V, Hur H, Liang Y, Nectow AR, Pillow JW, Witten IB (2017) Combined social and spatial coding in a descending projection from the prefrontal cortex. Cell 171:1663–1677.e1616
Okuyama T, Kitamura T, Roy DS, Itohara S, Tonegawa S (2016) Ventral CA1 neurons store social memory. Science 353:1536–1541
Parthasarathy A, Herikstad R, Bong JH, Medina FS, Libedinsky C, Yen S-C (2017) Mixed selectivity morphs population codes in prefrontal cortex. Nat Neurosci 20:1770–1779
Rao RP, von Heimendahl M, Bahr V, Brecht M (2019) Neuronal responses to conspecifics in the ventral CA1. Cell Rep 27:3460–3472.e3463
Rigotti M, Barak O, Warden MR, Wang X-J, Daw ND, Miller EK, Fusi S (2013) The importance of mixed selectivity in complex cognitive tasks. Nature 497:585
Wang F, Zhu J, Zhu H, Zhang Q, Lin Z, Hu H (2011) Bidirectional control of social hierarchy by synaptic efficacy in medial prefrontal cortex. Science 334:693–697
Zhan Y (2015) Theta frequency prefrontal–hippocampal driving relationship during free exploration in mice. Neuroscience 300:554–565
Zhou T, Zhu H, Fan Z, Wang F, Chen Y, Liang H, Yang Z, Zhang L, Lin L, Zhan Y, Wang Z, Hu H (2017) History of winning remodels thalamo-PFC circuit to reinforce social dominance. Science 357:162–168

Neural Circuit Mechanisms That Underlie Parental Care

6

Zi-Xian Yu, Xing-Yu Li, and Xiao-Hong Xu

Abstract

In mammals, parental care is essential for the survival of the young; therefore, it is vitally important to the propagation of the species. These behaviors, differing between the two sexes, are innate, stereotyped, and are also modified by an individual's reproductive experience. These characteristics suggest that neural mechanisms underlying parental behaviors are genetically hardwired, evolutionarily conserved as well as sexually differentiated and malleable to experiential changes. Classical lesion studies on neural control of parental behaviors, mostly done in rats, date back to the 1950s. Recent developments of new methods and tools in neuroscience, which allow precise targeting and activation/inhibition of specific populations of neurons and their projections to different brain structures, have afforded fresh opportunities to dissect and delineate the detailed neural circuit mechanisms that govern distinct components of parental behaviors in the genetically tractably organism, the laboratory mouse (*Mus musculus*). In this review, we summarize recent discoveries using modern neurobiological tools within the context of traditional lesion studies. In addition, we discuss interesting cross talk between neural circuits that govern parent care with those that regulate other innate behaviors such as feeding and mating.

Keywords

MPOA · Esr1 · Dopamine

6.1 Introduction

Parental behavior can be defined as any behavior carried out by a member of a species toward an immature conspecific that facilitate the survival of young and promote their well-being (Numan and Insel 2003a; Dulac et al. 2014). It can be further classified, according to the sex of the executor, into maternal behavior by females and paternal behavior by males. In most mammalian species, as females lactate they undertake most of the responsibilities to care for the young, while

Z.-X. Yu · X.-Y. Li
Institute of Neuroscience, State Key Laboratory of Neuroscience, CAS Center for Excellence in Brain Science and Intelligence Technology, Chinese Academy of Sciences, Shanghai, China

University of the Chinese Academy of Sciences, Beijing, China

Shanghai Center for Brain Science and Brain-Inspired Intelligence Technology, Shanghai, China

X.-H. Xu (✉)
Institute of Neuroscience, State Key Laboratory of Neuroscience, CAS Center for Excellence in Brain Science and Intelligence Technology, Chinese Academy of Sciences, Shanghai, China

Shanghai Center for Brain Science and Brain-Inspired Intelligence Technology, Shanghai, China
e-mail: xiaohong.xu@ion.ac.cn

H. Wang (ed.), *Neural Circuits of Innate Behaviors*, Advances in Experimental Medicine and Biology 1284, https://doi.org/10.1007/978-981-15-7086-5_6

bi-parental care, in which both males and females take parts in raising the offspring, occurs only in about 5–10% of mammalian species (Lonstein and De Vries 2000; Lukas and Huchard 2014; Kleiman and Malcolm 1981). Here, we focus our discussion to rats and mice as the majority of studies on the neurobiology of parental behaviors are carried out in these two species.

Parental behaviors in rodents consist of multiple stereotyped yet coordinated motor patterns that are often sequentially displayed and that are accompanied by suppression of other activities such as feeding and mating. Broadly speaking, maternal behaviors typically include several distinct pup-related behaviors (Numan and Insel 2003b; Hedrich 2013), including the following: (1) licking/grooming, which serve to clean the pups' body and stimulate defecation; (2) nest-building, building and maintaining a nest that is ~2–3 times larger and more completely enclosed than a common nest to provide a more comfortable environment for pups and better shield them from intruders; (3) pup retrieval, retrieval of pups that have strayed away back to the nest; (4) crouching, huddling over pups to ensures their warmth as they are still underdeveloped for proper thermoregulation; (5) the arched-back lactation/nursing posture. In addition, during the final days of pregnancy and throughout lactation, females will readily defend the nest and vigorously attacking the intruders, a behavior termed maternal aggression (Numan and Insel 2003b; Lonstein and Gammie 2002). The intensity of maternal aggression depends on litter size (Maestripieri and Alleva 2010) as well as food availability (Maestripieri 1991), suggesting a strict relationship to pup defense thus differentiating it from other forms of aggressive behaviors such as territorial aggression.

By comparison, virgin males are negligent toward pups and sometimes even attack and kill them (infanticide). Strikingly, male mice switch to become paternal following mating and cohabitation with a female, roughly around the time when its offspring is about to be born (Numan and Insel 2003a; Brown 1993; Tachikawa et al. 2013). This transition from infanticide to paternal care is thought to serve as an adaptive mechanism to prevent a male mouse from killing its own pups while still maintaining the ability to eliminate pups sired by other males (Elwood 1977). More importantly, when male mice do care for the young in the role of a father, their behaviors toward pups are very similar to those of females with some quantitative differences in extents and the fact that they do not lactate (Numan and Insel 2003a; Tachikawa et al. 2013; Elwood 1977). This indicates that maternal and paternal behaviors likely share some common neural mechanisms at a very basic level.

6.2 Factors That Influence the Display of Parental Behaviors

Parental behaviors are sensitively modulated by external cues from pups and by internal state that signals an individual's reproductive experience (Numan and Insel 2003a, b; Brown 1993). Therefore, before discussing brain areas and neural circuits that govern parental behaviors, it helps to first review these external and internal factors that influence the display of parental behaviors.

6.2.1 Olfactory and Auditory Cues Emitted by Pups

As rodents are nocturnal animals living in dark tunnels, females mainly locate pups through olfactory and auditory cues. A dam (mother) uses olfactory cues in milk and urine to establish a unique identity for her litters. Interfering with this olfactory identity by adding unwanted scents risks disrupting maternal behavior, particularly during the first week after parturition (Wilkinson and Miller 2010; Weber and Olsson 2008). In addition, pups emit a variety of ultrasonic vocalizations (USVs) when isolated from the dam or when they are away from the nest (Branchi et al. 2015). These USVs are crucial to direct the dams' attention and to trigger a maternal response (Ehret 2005; Ehret and Bernecker 1986). Manipulation of either the female's hearing ability (Ehret and Bernecker 1986;

Cohen-Salmon et al. 1985) or pups' vocal activity (Hood et al. 2010) decreases maternal responses and increases the latency for pup retrieval. Litter size (Priestnall 1972) as well as litters' sex composition (Paul 1991) also affect the duration of maternal caring. Females that rear large litters spend less time in the nest than females rearing small litters (Priestnall 1972). Similarly, females that nurse mixed litters (1 male + 3 female) spend more time on maternal behaviors and wean litters later than females that nurse all-male litters (Paul 1991).

6.2.2 Environmental Stressors

Moreover, maternal care is also influenced by environmental stressors in mice (Numan and Insel 2003a; Weber and Olsson 2008). Stressors of different intensity levels and kinds, including food and/or water deprivation, cage tilt, light, high temperature, and loud noise, have varying effects on the female's behavior toward pups with some stressor decreasing pup grooming and nursing and others eliciting infanticide and cannibalism (Meek et al. 2001; Macrì and Würbel 2007; Pardon et al. 2000).

6.2.3 Reproductive Status

Parental behaviors are also strongly modulated by an individual's parous or sexual experience. Nulliparous or virgin female rats avoid neonatal pups during the first encounter but can be induced to display maternal care after ~4–7 days of continuous exposure to pups (Numan and Insel 2003a; Weber and Olsson 2008) or by treatment of hormone regimens that are characteristic of parturient females (Siegel et al. 1978; Pedersen et al. 1982; Rosenblatt 1967). This process, called "sensitization" of maternal behaviors, is much quicker in virgin female mice (Dulac et al. 2014; Rosenblatt 1967). As mentioned before, virgin male mice, which typically neglect or attack pups (Elwood 1977), shift to paternal care after mating and/or lengthy cohabitation with a pregnant female (Numan and Insel 2003b; Brown 1993), indicating a slow experience-dependent "awakening" of paternal behaviors in males.

6.3 Brain Areas Involved in Parental Care

Given the complex behavioral repertoire of parental care and the multiple internal and external factors that influence its expression, it is not surprising that extensive brain areas/regions have been implicated in its regulation. These areas include olfactory and auditory system, regions of the hypothalamus, ventral tegmental area (VTA) and nucleus accumbens (NAc), periductal gray (PAG), and the medial prefrontal cortex (mPFC). These brain regions are highly interconnected and are thought to mediate the aspects of sensory processing, neural integration, motivation, motor output, and high cognitive control in the regulation of parental behaviors. Below, we will discuss the distinctive role played by each region with an emphasis on recent developments using cutting-edge tools (for review of these tools, please see references (Adamantidis et al. 2015; Atasoy and Sternson 2018; Zha and Xu 2015; Sternson et al. 2016)).

6.3.1 Main Olfactory Epithelium & Vomeronasal Organ (MOE & VNO)

In rodents, odors are detected by specialized sensory neurons that reside in the main olfactory epithelium (MOE) and the vomeronasal organ (VNO), which further relay the information to the main olfactory bulb (MOB) and the accessory olfactory bulb (AOB), respectively (Levy and Keller 2009). Olfactory cues play facilitatory roles in maternal behaviors but are not absolutely necessary for the initiation of the behavior in rats as grossly normal onset is found in primiparous females following destruction of either the MOE or VNO (Jirik-Babb et al. 1984; Fleming et al. 1992; Kolunie and Stern 1995). While in mice olfaction seems to play a more essential role in maternal behaviors. Irrigation of zinc sulfate within the MOE, which renders animals anosmic, results in the majority of the treated females eating their offsprings (Seegal and Denenberg 1974). Moreover, deficits in pup retrieval and maternal aggression are also observed in female mice null for *Cnga2* and *Trpc2*, two obligatory channels for odor-evoked neural activities in the MOE and VNO, respectively, demonstrating that maternal behaviors in mice require normal MOE and VNO function (Fraser and Shah 2014). On the other hand, detection of pup cues by VNO appears to promote pup attack behaviors in virgin males as surgical removal of the VNO or loss of VNO function in Trpc2 knockouts results in dramatic suppression of infanticide and even emergence of paternal behaviors in some males (Tachikawa et al. 2013; Mennella and Moltz 1988; Wu et al. 2014).

6.3.2 Auditory Cortex

Detection of pup USV calls is lateralized to the left auditory cortex in rodents (Marlin et al. 2015), which mirrors lateralization of language processing to the left hemisphere in humans (Loring et al. 1990; Bishop 2013). Functionally, infusion of the GABAA receptor agonist (muscimol) into the left auditory cortex impairs pup retrieval in females (Marlin et al. 2015). Interestingly, neural responses to pup calls in the auditory cortex undergo significant changes after parturition with higher neural correlation and higher signal-to-noise ratio observed in dams compared to pup-naive adult females (Gideon et al. 2013). Such changes are thought to better represent and transmit pup call information and the behavioral saliency of the stimuli (Gideon et al. 2013; Lior et al. 2011). Multisensory interaction is also reported as exposure to pups' odor reshape neuronal responses in the auditory cortex (Lior et al. 2011).

Curiously, Marlin et al. (2015) find that the left cortex expresses more oxytocin receptors than the right auditory cortex. Moreover, the display of pup retrieval is accelerated in virgin females following oxytocin injection into the left auditory cortex or stimulation of oxytocin neurons. Pairing pup calls with oxytocin delivery or activation of oxytocinergic fibers in the auditory cortex modulates population neural responses by acutely reducing inhibition but at the same time increasing the temporal correlation of inhibition and excitation over minutes to hours, thereby balancing the magnitude and timing of excitation and inhibition. Thus, oxytocin-induced plasticity in the auditory cortex likely underlie sensitization of behavioral response to pup calls in dams.

6.3.3 Medial Amygdala (MeA)

The amygdala represents a major brain region that processes olfactory information downstream of the MOB and the AOB (Sosulski et al. 2011; Ben-Shaul et al. 2010; Bergan et al. 2014). The AOB and MeA are more activated by pup interaction in virgin males than in fathers, indicating a role in infant aversion (Tachikawa et al. 2013; Li et al. 2017). Indeed, bilateral lesions of the MeA or unilateral lesion of the MeA paired with unilateral lesion of the anterior/ventromedial nuclei in the hypothalamus facilitate the expression of maternal behaviors in virgin female rats (Fleming et al. 1980; Sheehan et al. 2001), leading to the view that the MeA prevents the expression of maternal behavior through activation of hypothalamic nuclei. This view is further expanded by recent studies in mice showing that neurons in the posterodorsal division of the MeA (MeApd) encode pup cues and that such representation is potentiated by sexual experience in virgin animals of both sexes (Bergan et al. 2014; Li et al. 2017). Furthermore, stimulation of MeApd GABAergic neurons promotes pup grooming in virgin females, while inhibition of these neurons suppresses pup grooming (Chen et al. 2019). Interestingly, stimulation of the same population of MeA neurons in virgin males results in activity-level-dependent behavioral outputs: promoting pup grooming at low intensity and infanticide at high intensity (Chen et al. 2019). Together, these results show that the MeApd regulates paternal behaviors in a sex and experience-dependent manner.

6.3.4 Medial Preoptic Area (mPOA)

The medial preoptic area (mPOA) is a small brain region at the anterior tip of the hypothalamus and is perhaps the best studied brain region in the control of parental behaviors. The mPOA receives olfactory information via inputs from the MeA and from the bed nucleus of the stria terminalis (BNST) (Kohl et al. 2017; Kohl et al. 2018). Lesion or pharmacological inhibition of the mPOA blocks all consummatory components of maternal behaviors with no pup retrieval, nest-building, crouching or maternal aggression observed yet with pup contact remaining intact (Lee et al. 1999; Numan et al. 1988; Numan 1974; Arrati et al. 2006), which demonstrates that the mPOA is absolutely required for the initiation and execution of the consummatory aspects of maternal behaviors. Furthermore, lesion of the central part of the mPOA switch fathers from paternal care to infanticide, while activation of the same region attenuates infanticide in virgin males (Tsuneoka et al. 2015), suggesting an active role for this region to inhibit infanticide in fathers.

Using single cell profiling methods, it is recently revealed that the mPOA is an extremely heterogeneous structure consisting of many distinct neuronal subtypes which express different molecular markers such as neurotransmitters and hormone receptors (Kohl et al. 2018; Simerly et al. 1986; Tsuneoka et al. 2013; Tsuneoka et al. 2017). Along the same line, it is known that the mPOA regulates homeostatic processes such as thermoregulation (Szymusiak and Satinoff 1982), sleep (Chung et al. 2017) as well as parental behaviors. So the important question becomes what the identities of the neuronal populations that regulate parental behaviors are. Great progress has been made on this front using cell-type-specific tools that provide much improved spatial and temporal resolution on the function of defined neuronal populations than previous lesion and electrical stimulation methods (Wu et al. 2014; Kohl et al. 2018; Wei et al. 2018; Fang et al. 2018).

6.3.4.1 mPOA Galanin+ Neurons

Dr. Catherine Dulac's group of Harvard pioneered the field to explore the role of genetically defined populations of mPOA neurons in parental behaviors (Wu et al. 2014; Kohl et al. 2018). By screening various genetic markers, they first find that the neuropeptide galanin is enriched in mPOA neurons that are highly activated during parental behaviors. Via fiber photometry recordings of Ca^{2+} transient signals, they find that mPOA galanin+ ($mPOA^{galanin}$) neurons are highly activated during pup-directed behaviors (sniff, retrieval) but not during non-pup-directed behaviors (nest-building, crouching) or during social interactions with other adults in females or fathers (Kohl et al. 2018). More importantly, ablation of $mPOA^{galanin}$ neurons impairs all components of parental care and renders the ablated animals more likely to ignore or attack pups regardless of their sex or reproductive status, while ablation of a nearby population of mPOA neurons that express tyrosine hydroxylase (TH) has no effects. Furthermore, optogenetic activation of $mPOA^{galanin}$ neurons switches virgin male mice from pup avoidance/attack to pup grooming. Together, these results show that activation of $mPOA^{galanin}$ neurons suppresses aggressive behaviors toward pups and promotes pup grooming. However, as stimulation of $mPOA^{galanin}$ neurons fails to elicit other maternal behaviors such as pup retrieval and nest-building, it is likely that other populations of mPOA neurons play more prominent roles in these behaviors.

6.3.4.2 mPOA Esr1+ Neurons

Indeed, two independent studies recently point to the population of mPOA neurons that express estrogen receptor α (*Esr1*) as the vital player in controlling pup retrieval behavior (Wei et al. 2018; Fang et al. 2018). *Esr1* is expressed in about ~50% of mPOA neurons and is the canonical nuclear receptor for estrogen, which upon binding of estrogen translocates from the plasma to the nucleus to regulate multiple gene transcription (Wei et al. 2018). Classical studies show that estrogen injection intraperitoneally or directly into the mPOA facilitates the onset of maternal behaviors (Mayer et al. 1990; Rosenblatt and Ceus 1998; Siegel and Rosenblatt 1975), implying a function for mPOA^{Esr1} neurons in parental behaviors. Noteworthy, a subset of mPOA Esr1+ (mPOA^{Esr1}) neurons (15–30%), higher in males, co-express galanin and account for about half of the mPOA^{galanin} neurons (Wei et al. 2018).

Recently, via fiber photometry recording of Ca^{2+} transients, Wei et al. (2018) reveal ramping neural activates in mPOA^{Esr1} neurons prior to the initiation of pup retrieval behavior. This finding is independently validated by Fang et al. (2018), using single-unit recording of these neurons. Interestingly, this pattern of neural activity is specific to pup retrieval and is not observed in other components of maternal behaviors such as nest-building. Optogenetic activation of mPOA^{Esr1} neurons elicits pup retrieval in virgin animals of both sexes and even promotes retrieval or gathering of fake pups (Wei et al. 2018). Moreover, time-locked optogenetic inhibition of mPOA^{Esr1} neurons during pup contact significantly decreases the initiation of pup retrieval, while optogenetic inhibition after the initiation of a pup retrieval bout reduces the rate that the pup is successfully retrieved to the nest without affecting the duration of the retrieval behavior (Wei et al. 2018; Fang et al. 2018), indicating mPOA^{Esr1} neurons regulate not only the motor but also the "goal-directness" aspect of pup retrieval behavior.

6.3.4.3 mPOA Vgat+ Neurons

The majority (~80%) of mPOA neurons are inhibitory and express the vesicular GABA transporter (Vgat) as the marker, of which about half is Esr1+ (Wei et al. 2018). As mPOA^{galanin} and mPOA^{Esr1} neurons seem to regulate pup grooming and pup retrieval, respectively, is there a specific population of mPOA neurons that regulate maternal nest-building? Single-unit recording of mPOA neurons show that ~20% mPOA neurons are activated during pup retrieval while ~5% are activated during nest-building (Fang et al. 2018). Moreover, neurons activated during retrieval tend to be inhibited during nest-building (Fang et al. 2018). Interestingly, Li et al. (2019) show that optogenetic activation of mPOA Vgat+ (mPOA^{Vgat}) neurons elicits both pup retrieval and nest-building, while stimulation of mPOA^{Esr1} neurons elicits only pup retrieval. Consistently, behavioral-locked optogenetic inhibition of mPOA^{Vgat} neurons disrupts both pup retrieval and maternal nest-building. Thus, it seems that a distinct subset of mPOA^{Vgat} neurons that does not express Esr1 may underlie maternal nest-building. This hypothesis remains to be tested more directly.

6.3.5 Ventral Tegmental Area (VTA)

One of the major downstream targets of the mPOA outside of the hypothalamus is the ventral tegmental area (VTA), which sends out dopaminergic projections throughout the brain to regions such as the nucleus accumbens (NAc) and is critically involved in motivation and reinforcement learning (Salamone and Correa 2012; McHenry et al. 2017). Pharmacological inactivation of the VTA reduces pup licking and retrieval in postpartum females and blocks the expression of place preference for pup-paired context (Seip and Morrell 2009; Numan and Smith 1984; Keer and Stern 1999), while destruction of VTA dopamine neurons or depletion of dopamine terminals in the ventral striatum also causes a persistent deficit in pup retrieval (Hansen et al. 1991a, b). Similarly, lesion of the shell but not the core of the NAc significantly disrupts pup retrieval behavior without affecting other components of maternal behavior such as pup licking, nest-building, and nursing (Li and Fleming 2003), while infusion dopamine receptor antagonist in NAc inhibits pup retrieval and pup licking but enhances nursing in lactating rats (Keer and Stern 1999).

Via fiber photometry recordings of Ca^{2+} transients, Fang et al. (2018) find that VTA dopamine neurons are acutely and strongly activated during pup retrieval and that optogenetic stimulation of mPOA^{Esr1} → VTA projections promotes pup retrieval behavior. Consistent with the vast of majority mPOA^{Esr1} being GABAergic, they further show that mPOA^{Esr1} neurons send strong inhibitory inputs preferentially to VTA non-dopaminergic cells, which locally inhibit VTA dopamine neurons. Thus, stimulation of mPOA^{Esr1} neurons would result in a net activation of VTA dopamine neurons via a dis-inhibitory mechanism. Indeed, pharmacological inhibition of the VTA blocks pup retrieval behavior elicited by activation of mPOA^{Esr1} neurons. Similarly, mPOA^{galanin} neurons also project to the VTA. Activation or inhibition of mPOA^{galanin} → VTA projections bidirectionally change the animal's motivation to interact with pups as measured by the number of times that the animal crosses a barrier that separates it from the pups (Kohl et al. 2018).

6.3.6 Periaqueductal Gray (PAG)

The periaqueductal gray (PAG) is composed of several distinct longitudinal neuronal columns and is thought to convey motor outputs of various innate behaviors including parental care (Zha and Xu 2015; Deng et al. 2016; Watson et al. 2016). Paradoxically, cytotoxic lesions of PAG facilitates maternal response (Sukikara et al. 2010), while injection of $GABA_A$ receptor antagonist into the PAG dose-dependently promotes pup grooming but impairs maternal aggression in lactating females (Lee and Gammie 2010). Indeed, mPOA^{galanin} neurons send strong projection to the PAG and preferentially synapse onto PAG GABAergic neurons (Kohl et al. 2018). Moreover, optogenetic activation of mPOA^{galanin} → PAG projections suppresses pup attack in virgin males and promotes pup grooming and pup-directed sniffing in both males and females without affecting other parental behaviors or affecting the parental motivation to cross a barrier to interact with pups. Furthermore, optogenetic inhibition of mPOA^{galanin} → PAG projections significantly reduces pup grooming and pup-directed sniffing without affecting other behaviors. Given that ~90% of mPOA^{galanin} neurons are GABAergic and that mPOA^{galanin} neurons preferentially target PAG GABAergic neurons, it is possible that stimulation of the mPOA may recruit different subsets of PAG excitatory neurons via a dis-inhibitory mechanism; however, the detailed cellular mechanism through which mPOA → PAG projections coordinate different components of parental behaviors remains to be investigated.

6.3.7 Paraventricular Nucleus of the Hypothalamus (PVN)

The mPOA also project to the paraventricular nucleus of the hypothalamus (PVN). Electrolytic lesions of the PVN on day 15 of gestation in rats disrupts nearly all maternal behaviors while lesion performed on day 4 postpartum has little effects, suggesting that PVN specifically regulates the initiation but not the maintenance of maternal behavior in rats (Insel and Harbaugh 1989). Oxytocin, a neuropeptide critical for parturition and lactation, is synthesized in a subset of neurons in the PVN (Richard et al. 1991; Gimpl and Fahrenholz 2001). Intraperitoneal injection of oxytocin or infusion of oxytocin into brain regions such as the auditory cortex, mPOA, or VTA accelerates the expression of maternal behaviors in virgin females (Pedersen et al. 1982; Marlin et al. 2015; Pedersen et al. 1994), while infusions of an oxytocin receptor antagonist into the VTA or the mPOA blocks pup retrieval and nursing postures in dams (Pedersen et al. 1994). These results support that PVN oxytocin (PVN^{OXT}) neurons regulate the initiation of maternal behavior (Insel and Harbaugh 1989). Indeed, optogenetic stimulation of PVN^{OXT} neurons or their projections in the auditory cortex accelerates pup retrieval and decreases retrieval latency in virgin females (Marlin et al. 2015). Interestingly, Scott et al. (2015) identify a sexually dimorphic projection (more prominent in females than males) from TH-expressing neurons in the anterior periventricular nucleus (AVPe) to PVN^{OXT} neurons that modulate oxytocin secretion. Ablation of AVPe TH+ (AVPeTH) neurons decreases oxytocin level and disrupts pup retrieval along with other maternal behaviors, while optogenetic stimulation of these neuron positively promotes maternal behaviors. Interestingly, AVPeTH neurons do not appear to regulate paternal behaviors in males but rather act to suppress aggression toward other males, suggesting a sexually dimorphic function.

6.3.8 Medial Prefrontal Cortex (mPFC)

While the medial prefrontal cortex (mPFC) is thought to control high cognitive functions, several studies have also reported its role in maternal behaviors. By c-Fos staining, increased neural activities are observed in the infralimbic area of the mPFC in maternal rats that develop place preference for pup-paired context (Mattson and Morrell 2005). Rodent neuroimaging studies have also reported increases in blood oxygen level-dependent (BOLD) signals in the mPFC in lactating rats in response to suckling stimulation from pups (Febo and Ferris 2007; Marcelo et al. 2005). In addition, functional MRI (fMRI) studies have also implicated the mPFC in maternal care in humans (Bartels and Zeki 2004; Ranote et al. 2004; Lane et al. 2008). Moreover, excitotoxic lesion of the mPFC prior to pregnancy impairs some maternal behaviors such as pup retrieval and pup licking but spare other behaviors such as nest-building (Afonso et al. 2007), while tetrodotoxin (TTX)-mediated inactivation or GABA-mediated inhibition of the mPFC leads to dramatic reduction in pup retrieval in rats (Febo et al. 2010). Together, these results show that the mPFC also participates in the regulation of parental care.

6.3.9 Ventrolateral Division of the Ventromedial Hypothalamus (VMHvl)

While the above sections focus on pup-directed aspects of parental behaviors, some progress has also been made on the neural control of maternal aggression. Specifically, Esr1+ neurons in the ventrolateral division of the ventromedial hypothalamus (VMHvl), a population known to control male territorial aggression and female sexual behaviors (Falkner et al. 2016; Lee et al. 2014; Yang et al. 2013; Lin et al. 2011), are recently shown to be activated when lactating females attack stranger intruders (Hashikawa et al. 2017). Furthermore, inactivation of VHMvl^{Esr1} cells reduces maternal aggression, whereas activation of these neurons elicits attack in virgin females (Hashikawa et al. 2017). Interestingly, two seemingly topographically separable subdivisions of VMHvl^{Esr1} neurons, which differ in gene expression and projection patterns, may distinctively regulate female sexual behaviors versus maternal aggression. In particular, the more medially located division of VMHvl^{Esr1} neurons, which preferentially projects to the PAG, is more activated during maternal aggression, whereas the other more laterally located division that projects to both AVPV and PAG is more activated during female sexual behavior (Hashikawa et al. 2017).

6.4 Cross Talk Between Neural Control of Parental Care and Other Innate Behaviors

Ethologists have long noted hierarchical and antagonistic control of behaviors. Indeed, neural control of parental behaviors must be coordinated with neural control of other behaviors such as mating and aggression to achieve a balance among different behaviors and to maximize fitness. This is most likely achieved via lateral interactions between neural structures that regulate each behavior at many different levels.

6.4.1 Overlaps Between Neural Control of Paternal Care and Other Social Behaviors

It is found that ablation of mPOA^{galanin} neurons results in not only marked impairments of parental responses but also in defects in mating behaviors in males (Wu et al. 2014). Along the same line, optogenetic activation of mPOA^{galanin} neurons while promoting pup grooming and suppressing pup-directed aggression also suppresses inter-male territorial aggression

(Wu et al. 2014), indicating the complex roles that $mPOA^{galanin}$ neurons play in pup-directed behaviors versus social behaviors with other adult conspecifics. Interestingly, a subset of $mPOA^{galanin}$ neurons specifically project to the MeA and are broadly activated during parental behaviors. Optogenetic activation of this $mPOA^{galanin} \rightarrow$ MeA projection has little effects on parental behaviors but rather inhibits inter-male aggression and decreases the amount of time that a female spent chemo-investigating a male intruder (Kohl et al. 2018), suggesting that this pathway may function to inhibit interactions with adult conspecifics during parental behaviors.

Similarly, $mPOA^{Esr1}$ neurons, the population that is both necessary and sufficient for pup retrieval behavior in both sexes, also regulate male mating behavior. Wei et al. (Wei et al. 2018) find that $mPOA^{Esr1}$ neuron activities ramped before the onset of pup retrieval as well as male mating. Furthermore, optogenetic activation of mPOA ^{Esr1} neurons elicits male-typical mating behaviors in both sexes when the mice are presented with a female intruder while ablation or optogenetic inhibition of $mPOA^{Esr1}$ disrupts mating behavior in males (Wei et al. 2018). It is further shown that neurons activated in response to pups or females are not the same but are two separable subpopulations that overlap. Together, these results reveal the shared layout within which $mPOA^{Esr1}$ neurons function to broadly regulate sexually dimorphic behaviors in both male and female mice.

6.4.2 Reciprocal Antagonisms Between Neural Control of Maternal Care and Feeding

In species such as mouth-breeding cichlid fish or domestic chicken, females endure lengthy voluntary anorexia during brood care, spending time sitting in the nest or caring for the offspring instead of feeding or food foraging (Mrowka 1986; Mrosovsky and Sherry 1980). Similar phenomenon is also observed in rodents where mothers spend almost all her time in the early postpartum period curling around the pups in the nest and are not seen resting alone without body contacts to pups until day 9 postpartum (Numan and Insel 2003b; Konig and Markl 1987). Recently, Han et al. (2017) find that the presence of pups strongly delays and decreases food consumption in physiologically fasted virgin female and male mice and even when Agrp neurons in the arcuate nucleus (ARC^{Agrp}), the quintessential hunger neurons, are optogenetically activated. Furthermore, chemogenetic activation of Vglut2+ but not Vgat+ neurons in the mPOA is sufficient to suppress hunger-induced feeding. Meanwhile, Li et al. (2019) find that ARC^{Agrp} neurons form inhibitory synapses onto ~30% of $mPOA^{Vgat}$ neurons, and activation of these projections in females dramatically inhibits maternal nest-building without affecting pup retrieval behavior. Together, these data support a model that hunger-induced activation of ARC^{Agrp} neurons inhibits mPOA Vgat to suppress maternal nest-building, whereas in parallel activation of mPOA Vglut2 neurons by pup-derived cues delays feedings. Such reciprocal antagonism between hunger and maternal care may allow a female to better balance feeding with different components of pup care and to prioritize behaviors that are more urgent.

6.5 Toward a Neural Circuit Mechanism for Parental Care

Brain regions and neural pathways discussed here do not function in isolation but rather form a large interconnected network that integrate external pup-derived sensory cues with internal hormonal factors and experience-induced plasticity to control discrete components of parental care such as pup grooming, pup retrieval, nest-building, and maternal aggression as well as the motivation to interact with pups, within the frame of a much larger neural network that controls and coordinates all behaviors. Thus, behavioral control is not instantiated by a single group of neurons but is mediated by concerted neural activities distributed across the entire neural network. Nevertheless, identification of genetic defined populations of neurons provides the essential entry points to delineate the architecture and the logic of the underlying neural circuit.

Using pseudorabies virus, Kohl et al. (2018) trans-synaptically labeled presynaptic inputs to $mPOA^{galanin}$ neurons and identified >20 brain regions including MeA, BNST, ARC, VMH, and PVN. Interestingly, ~20% presynaptic inputs originate within the mPOA, indicating extensive local processing, while another ~40% of presynaptic inputs are from other hypothalamic regions, of which the ARC provides the most abundant inputs. By combining retrograde tracing with c-Fos staining, the authors show that the majority of these upstream regions are activated during parental behaviors, albeit in a sex and reproductive state-dependent manner. For instance, while local mPOA inputs are activated during parent behaviors in virgin females, mothers, and fathers, more activation of neurons in the pheromone-processing pathway (MeA and BNST) is observed in fathers and virgin females but not mothers, suggesting that the MeA-BNST pathway, which mediates pup-directed aggression, is silenced in mothers but remains partially active in sexually experienced males and parental virgin females.

Next, by tracing neuronal projection patterns, Kohl et al. (2018) further show that $mPOA^{galanin}$ neurons project to >20 brain regions, many of which overlap with the regions that provide presynaptic inputs to these neurons, indicating extensive reciprocal connectivity within the circuit. Inside the hypothalamus, $mPOA^{galanin}$ neurons densely project to the PVN and AVPe, while outside of the hypothalamus they project to the VTA, PAG, and MeA along with other regions. $mPOA^{galanin}$ neurons that project to a given target show minimum overlap, occupy anatomically distinctive positions within the mPOA, and receive broad but pathway-specific combinations of inputs that show characteristic enrichment or depletion patterns. For example, inputs from the NAc and lateral septum preferentially target VTA-projecting $mPOA^{galanin}$ neurons. In addition, there are great cellular specificity within the connectivity patterns. For instance, $mPOA^{galanin}$ neurons receive synaptic inputs from PVN arginine vasopressin (AVP) neurons (PVN^{AVP}) but not from PVN^{OXT} neurons. In return, $mPOA^{galanin}$ neurons project to PVN^{AVP}, PVN^{OXT}, and PVN corticotropin-releasing hormone (PVN^{CRH}) neurons, in a sexually dimorphic manner with more $mPOA^{galanin}$ neurons projecting to PVN^{AVP} and PVN^{CRH} neurons in males and more $mPOA^{galanin}$ neurons projecting to PVN^{OXT} neurons in females. Similarly, $mPOA^{galanin}$ neurons do not receive inputs from but send projections to $AVPe^{TH}$ neurons, which further send a female-biased projection to PVN^{OXT} neurons, further supporting an intimate connection between the $mPOA^{galanin}$ neural circuit and sex-specific activation of PVN^{OXT} neurons.

Additionally, PAG, VTA, and MeA-projecting $mPOA^{galanin}$ neurons are tuned to different aspects of parental behaviors with the PAG-projecting population specifically activated during pup grooming, the MeA-projecting population broadly activated during nearly all pup-directed behaviors and the VTA-projecting population show minimum activity except some minor activation in a subset of females when they enter an empty nest. Functionally, optogenetic activation or inhibition of $mPOA^{galanin}$ → PAG projections promote or suppresses pup grooming, respectively, without affecting the parental motivation to interact with pups. By comparison, optogenetic activation $mPOA^{galanin}$ → VTA projections increase the number of times that the stimulated animals cross a climbable barrier to interact with pups, indicating increased parental motivation, without affecting the actual duration or quality of pup interactions such that a non-parental male bearing the $mPOA^{galanin}$ → VTA stimulation would actually cross the barrier to attack pups. Interestingly, activation $mPOA^{galanin}$ → MeA projections has little effects on either the motivation or motor aspects of parental behaviors, but rather suppress male–male aggression and the amount of time that a female investigate a male intruder, indicating inhibition of other competing social behaviors or interests. Taken together, these findings support a model in which largely non-overlapping, projection-defined $MPOA^{galanin}$ subpopulations form functional modules to integrate broad input combinations and regulate specific aspects of parental behaviors. Future studies are required to determine whether the function circuits that other genetic defined populations of neurons (such as $mPOA^{Esr1}$, $mPOA^{Vgat}$) act in to regulate parental behaviors are also organized in a similar manner.

6.6 Conclusion & Perspective

Parental care is essential for the survival of the young. While the mPOA represents a key node for the control of parental behaviors behavior, the neural network that it functions in is distributed. Great progress has been made on the neural circuit mechanism that govern parental behaviors, aided mainly by recent tool developments that allow cell-type-specific monitoring and manipulation of neuronal activities. Nevertheless, outstanding questions remain. In particular, as neuronal subtypes can be defined according to gene expression, connectivity, and developmental profiles, the exact role that each neuronal type play in various components and aspects of paternal behaviors await more complete and comprehensive investigations. Built on such knowledge, it will be interesting to determine whether maternal or paternal behaviors, which appear similar from outside, rely on similar neuronal populations. Along this line, how are such neurons modified by experience, activated or suppressed by pup-derived cues in a sex and state-dependent manner, and how do they compete with neurons that govern other behaviors to select a particular motor output? In summary, it is an exciting time for the study of parental care, and many great questions await further explorations.

References

Adamantidis A et al (2015) Optogenetics: 10 years after ChR2 in neurons--views from the community. Nat Neurosci 18:1202–1212. https://doi.org/10.1038/nn.4106

Afonso V, Sison MV, Fleming A (2007) Medial prefrontal cortex lesions in the female rat affect sexual and maternal behavior and their sequential organization. Behav Neurosci 121:515–526

Arrati PG, Carmona C, Dominguez G, Beyer C, Rosenblatt JS (2006) GABA receptor agonists in the medial preoptic area and maternal behavior in lactating rats. Physiol Behav 87:51–65. https://doi.org/10.1016/j.physbeh.2005.08.048

Atasoy D, Sternson SM (2018) Chemogenetic tools for causal cellular and neuronal biology. Physiol Rev 98:391–418. https://doi.org/10.1152/physrev.00009.2017

Bartels A, Zeki S (2004) The neural correlates of maternal and romantic love. NeuroImage 21:1155–1166

Ben-Shaul Y, Katz LC, Mooney R, Dulac C (2010) In vivo vomeronasal stimulation reveals sensory encoding of conspecific and allospecific cues by the mouse accessory olfactory bulb. Proc Natl Acad Sci U S A 107:5172–5177. https://doi.org/10.1073/pnas.0915147107

Bergan JF, Ben-Shaul Y, Dulac C (2014) Sex-specific processing of social cues in the medial amygdala. elife 3:e02743. https://doi.org/10.7554/eLife.02743

Bishop DV (2013) Cerebral asymmetry and language development: cause, correlate, or consequence? Science (New York, NY) 340:1230531. https://doi.org/10.1126/science.1230531

Branchi I, Santucci D, Vitale A, Alleva E (2015) Ultrasonic vocalizations by infant laboratory mice: a preliminary spectrographic characterization under different conditions. Dev Psychobiol 33:249–256

Brown RE (1993) Hormonal and experiential factors influencing parental behaviour in male rodents: an integrative approach. Behav Process 30:1–27. https://doi.org/10.1016/0376-6357(93)90009-g

Chen PB et al (2019) Sexually dimorphic control of parenting behavior by the medial amygdala. Cell 176:1206–1221. e1218. https://doi.org/10.1016/j.cell.2019.01.024

Chung S et al (2017) Identification of preoptic sleep neurons using retrograde labelling and gene profiling. Nature 545:477. https://doi.org/10.1038/nature22350. https://www.nature.com/articles/nature22350#supplementary-information

Cohen-Salmon C et al (1985) Differences in patterns of pup care in mice. V--Pup ultrasonic emissions and pup care behavior. Physiol Behav 35:167–174

Deng H, Xiao X, Wang Z (2016) Periaqueductal gray neuronal activities underlie different aspects of defensive behaviors. J Neurosci 36:7580–7588. https://doi.org/10.1523/jneurosci.4425-15.2016

Dulac C, O'Connell LA, Wu Z (2014) Neural control of maternal and paternal behaviors. Science 345:765–770. https://doi.org/10.1126/science.1253291

Ehret G (2005) Infant rodent ultrasounds – a gate to the understanding of sound communication. Behav Genet 35:19–29

Ehret G, Bernecker C (1986) Low-frequency sound communication by mouse pups (Mus musculus): wriggling

calls release maternal behaviour. Anim Behav 34:821–830
Elwood RW (1977) Changes in the responses of male and female gerbils (Meriones unguiculatus) towards test pups during the pregnancy of the female. Anim Behav 25:46–51. https://doi.org/10.1016/0003-3472(77)90066-5
Falkner AL, Grosenick L, Davidson TJ, Deisseroth K, Lin D (2016) Hypothalamic control of male aggression-seeking behavior. Nat Neurosci 19:596–604. https://doi.org/10.1038/nn.4264
Fang YY, Yamaguchi T, Song SC, Tritsch NX, Lin D (2018) A hypothalamic midbrain pathway essential for driving maternal behaviors. Neuron 98:192–207.e110. https://doi.org/10.1016/j.neuron.2018.02.019
Febo M, Felix-Ortiz AC, Johnson TR (2010) Inactivation or inhibition of neuronal activity in the medial prefrontal cortex largely reduces pup retrieval and grouping in maternal rats. Brain Res 1325:77–88. https://doi.org/10.1016/j.brainres.2010.02.027
Febo M, Ferris CF (2007) Development of cocaine sensitization before pregnancy affects subsequent maternal retrieval of pups and prefrontal cortical activity during nursing. Neuroscience 148:400–412
Fleming AS, Gavarth K, Sarker J (1992) Effects of transections to the vomeronasal nerves or to the main olfactory bulbs on the initiation and long-term retention of maternal behavior in primiparous rats. Behav Neural Biol 57:177–188
Fleming AS, Vaccarino F, Luebke C (1980) Amygdaloid inhibition of maternal behavior in the nulliparous female rat. Physiol Behav 25:731–743. https://doi.org/10.1016/0031-9384(80)90377-7
Fraser EJ, Shah NM (2014) Complex chemosensory control of female reproductive behaviors. PLoS One 9: e90368. https://doi.org/10.1371/journal.pone.0090368
Gideon R, Lior C, Adi M, Israel N (2013) Elevated correlations in neuronal ensembles of mouse auditory cortex following parturition. J Neurosci 33:12851–12861
Gimpl G, Fahrenholz F (2001) The oxytocin receptor system: structure, function, and regulation. Physiol Rev 81:629–683. https://doi.org/10.1152/physrev.2001.81.2.629
Han Y, Li XY, Wang SR, Wei YC, Xu XH (2017) Presence of pups suppresses hunger-induced feeding in virgin adult mice of both sexes. Neuroscience 362:228–238. https://doi.org/10.1016/j.neuroscience.2017.08.047
Hansen S, Harthon C, Wallin E, Lã Fberg L, Svensson K (1991a) Mesotelencephalic dopamine system and reproductive behavior in the female rat: effects of ventral tegmental 6-hydroxydopamine lesions on maternal and sexual responsiveness. Behav Neurosci 105:588–598
Hansen S, Harthon C, Wallin E, Löfberg L, Svensson K (1991b) The effects of 6-OHDA-induced dopamine depletions in the ventral or dorsal striatum on maternal and sexual behavior in the female rat. Pharmacol Biochem Behav 39:71–77
Hashikawa K et al (2017) Esr1(+) cells in the ventromedial hypothalamus control female aggression. Nat Neurosci 20:1580–1590. https://doi.org/10.1038/nn.4644
Hedrich H (2013) The laboratory mouse (second edition). Anticancer Res 33:831–845
Hood KE, Dreschel NA, Granger DA (2010) Maternal behavior changes after immune challenge of neonates with developmental effects on adult social behavior. Dev Psychobiol 42:17–34
Insel TR, Harbaugh CR (1989) Lesions of the hypothalamic paraventricular nucleus disrupt the initiation of maternal behavior. Physiol Behav 45:1033–1041. https://doi.org/10.1016/0031-9384(89)90234-5
Jirik-Babb P, Manaker S, Tucker AM, Hofer MA (1984) The role of the accessory and main olfactory systems in maternal behavior of the primiparous rat. Behav Neural Biol 40:170–178
Keer SE, Stern JM (1999) Dopamine receptor blockade in the nucleus accumbens inhibits maternal retrieval and licking, but enhances nursing behavior in lactating rats. Physiol Behav 67:659–669. https://doi.org/10.1016/s0031-9384(99)00116-x
Kleiman DG, Malcolm JR (1981) The evolution of male parental investment in mammals. Q Rev Biol 52:39–68
Kohl J, Autry AE, Dulac C (2017) The neurobiology of parenting: a neural circuit perspective. BioEssays 39:1–11. https://doi.org/10.1002/bies.201600159
Kohl J et al (2018) Functional circuit architecture underlying parental behaviour. Nature 556:326–331. https://doi.org/10.1038/s41586-018-0027-0
Kolunie JM, Stern JM (1995) Maternal aggression in rats: effects of olfactory bulbectomy, ZnSO4-induced anosmia, and vomeronasal organ removal. Horm Behav 29:492–518. https://doi.org/10.1006/hbeh.1995.1285
Konig B, Markl H (1987) Maternal-care in house mice.1. The weaning strategy as a means for parental manipulation of offspring quality. Behav Ecol Sociobiol 20:1–9. https://doi.org/10.1007/Bf00292161
Lane S, Jian L, Peter FP, Read M (2008) What's in a smile? Maternal brain responses to infant facial cues. Pediatrics 122:40–51
Lee A, Clancy S, Fleming AS (1999) Mother rats bar-press for pups: effects of lesions of the mpoa and limbic sites on maternal behavior and operant responding for pup-reinforcement. Behav Brain Res 100:15–31
Lee G, Gammie SC (2010) GABA A receptor signaling in caudal periaqueductal gray regulates maternal aggression and maternal care in mice. Behav Brain Res 213:230–237
Lee H et al (2014) Scalable control of mounting and attack by Esr1+ neurons in the ventromedial hypothalamus. Nature 509:627–632. https://doi.org/10.1038/nature13169
Levy F, Keller M (2009) Olfactory mediation of maternal behavior in selected mammalian species. Behav Brain Res 200:336–345. https://doi.org/10.1016/j.bbr.2008.12.017

Li M, Fleming AS (2003) The nucleus accumbens shell is critical for normal expression of pup-retrieval in postpartum female rats. Behav Brain Res 145:99–111

Li XY et al (2019) AGRP neurons project to the medial preoptic area and modulate maternal nest-building. J Neurosci 39:456–471. https://doi.org/10.1523/jneurosci.0958-18.2018

Li Y et al (2017) Neuronal representation of social information in the medial amygdala of awake behaving mice. Cell 171:1176–1190 .e1117. https://doi.org/10.1016/j.cell.2017.10.015

Lin D et al (2011) Functional identification of an aggression locus in the mouse hypothalamus. Nature 470:221–226. https://doi.org/10.1038/nature09736

Lior C, Gideon R, Adi M (2011) Multisensory integration of natural odors and sounds in the auditory cortex. Neuron 72:357–369

Lonstein JS, De Vries GJ (2000) Sex differences in the parental behavior of rodents. Neurosci Biobehav Rev 24:669–686

Lonstein JS, Gammie SC (2002) Sensory, hormonal, and neural control of maternal aggression in laboratory rodents. Neurosci Biobehav Rev 26:869–888

Loring DW et al (1990) Cerebral language lateralization: evidence from intracarotid amobarbital testing. Neuropsychologia 28:831–838. https://doi.org/10.1016/0028-3932(90)90007-b

Lukas D, Huchard E (2014) Sexual conflict. The evolution of infanticide by males in mammalian societies. Science (New York, NY) 346:841–844. https://doi.org/10.1126/science.1257226

Macrì S, Würbel H (2007) Effects of variation in postnatal maternal environment on maternal behaviour and fear and stress responses in rats. Anim Behav 73:171–184

Maestripieri D (1991) Litter gender composition, food availability, and maternal defence of the young in house mice (Mus domesticus). Behaviour 116:239–251

Maestripieri D, Alleva E (2010) Maternal aggression and litter size in the female house mouse. Ethology 84:27–34

Marcelo F, Michael N, Ferris CF (2005) Functional magnetic resonance imaging shows oxytocin activates brain regions associated with mother-pup bonding during suckling. J Neurosci 25:11637–11644

Marlin BJ, Mitre M, D'Amour JA, Chao MV, Froemke RC (2015) Oxytocin enables maternal behaviour by balancing cortical inhibition. Nature 520:499–504. https://doi.org/10.1038/nature14402

Mattson BJ, Morrell JI (2005) Preference for cocaine-versus pup-associated cues differentially activates neurons expressing either Fos or cocaine- and amphetamine-regulated transcript in lactating, maternal rodents. Neuroscience 135:315–328

Mayer AD, Monroy MA, Rosenblatt JS (1990) Prolonged estrogen-progesterone treatment of nonpregnant ovariectomized rats: factors stimulating home-cage and maternal aggression and short-latency maternal behavior. Hormones & Behavior 24:342–364

McHenry JA et al (2017) Hormonal gain control of a medial preoptic area social reward circuit. Nat Neurosci 20:449. https://doi.org/10.1038/nn.4487. https://www.nature.com/articles/nn.4487#supplementary-information

Meek LR, Dittel PL, Sheehan MC, Chan JY, Kjolhaug SR (2001) Effects of stress during pregnancy on maternal behavior in mice. Physiol Behav 72:473–479

Mennella JA, Moltz H (1988) Infanticide in the male rat: the role of the vomeronasal organ. Physiol Behav 42:303–306. https://doi.org/10.1016/0031-9384(88)90087-x

Mrosovsky, N., . & Sherry, D. F. Animal anorexias. Science 207, 837 (1980)

Mrowka W (1986) Satiation restores brood care motivation in the female mouthbrooder Pseudocrenilabrus multicolor (Cichlidae). Physiol Behav 38:153–156

Numan M (1974) Medial preoptic area and maternal behavior in the female rat. J Comp Physiol Psychol 87:746–759

Numan M, Corodimas KP, Numan MJ, Factor EM, Piers WD (1988) Axon-sparing lesions of the preoptic region and substantia innominata disrupt maternal behavior in rats. Behav Neurosci 102:381

Numan M, Insel TR (2003a) The neurobiology of parental behavior, vol 1. Springer, New York

Numan, M. & Insel, T. R. (2003b) The neurobiology of parental behavior

Numan M, Smith HG (1984) Maternal behavior in rats: evidence for the involvement of preoptic projections to the ventral tegmental area. Behav Neurosci 98:712–727

Pardon M, Gerardin P, Joubert C, Perez-Diaz F, Cohen-Salmon C (2000) Influence of prepartum chronic ultramild stress on maternal pup care behavior in mice. Biol Psychiatry 47:858–863. https://doi.org/10.1016/s0006-3223(99)00253-x

Paul ES (1991) Litter composition affects parental care, offspring growth and the development of aggressive behaviour in wild house mice. Behaviour 116:90–108

Pedersen CA, Ascher JA, Monroe YL, Prange AJ (1982) Oxytocin induces maternal behavior in virgin female rats. Science 216:648–650

Pedersen CA, Caldwell JD, Walker C, Ayers G, Mason GA (1994) Oxytocin activates the postpartum onset of rat maternal behavior in the ventral tegmental and medial preoptic areas. Behav Neurosci 108:1163–1171

Priestnall R (1972) Effects of litter size on the behaviour of lactating female mice (Mus musculus). Anim Behav 20:386–394

Ranote, S., . et al. The neural basis of maternal responsiveness to infants: an fMRI study. Neuroreport 15, 1825–1829 (2004)

Richard P, Moos F, Freund-Mercier MJ (1991) Central effects of oxytocin. Physiol Rev 71:331–370. https://doi.org/10.1152/physrev.1991.71.2.331

Rosenblatt JS (1967) Nonhormonal basis of maternal behavior in the rat. Science (New York, NY)

156:1512–1514. https://doi.org/10.1126/science.156.3781.1512

Rosenblatt JS, Ceus K (1998) Estrogen implants in the medial preoptic area stimulate maternal behavior in male rats. Hormones & Behavior 33:23–30

Salamone JD, Correa M (2012) The mysterious motivational functions of mesolimbic dopamine. Neuron 76:470–485. https://doi.org/10.1016/j.neuron.2012.10.021

Scott N, Prigge M, Yizhar O, Kimchi T (2015) A sexually dimorphic hypothalamic circuit controls maternal care and oxytocin secretion. Nature 525:519–522. https://doi.org/10.1038/nature15378

Seegal RF, Denenberg VH (1974) Maternal experience prevents pup-killing in mice induced by peripheral anosmia. Physiol Behav 13:339–341. https://doi.org/10.1016/0031-9384(74)90056-0

Seip KM, Morrell JI (2009) Transient inactivation of the ventral tegmental area selectively disrupts the expression of conditioned place preference for pup- but not cocaine-paired contexts. Behav Neurosci 123:1325–1338

Sheehan T, Paul M, Amaral E, Numan MJ, Numan M (2001) Evidence that the medial amygdala projects to the anterior/ventromedial hypothalamic nuclei to inhibit maternal behavior in rats. Neuroscience 106:341–356

Siegel HI, Doerr HK, Rosenblatt JS (1978) Further studies on estrogen-induced maternal behavior in hysterectomized-ovariectomized virgin rats. Physiol Behav 21:99–103

Siegel HI, Rosenblatt JS (1975) Hormonal basis of hysterectomy-induced maternal behavior during pregnancy in the rat. Hormones & Behavior 6:211–222

Simerly RB, Gorski RA, Swanson LW (1986) Neurotransmitter specificity of cells and fibers in the medial preoptic nucleus: an immunohistochentical study in the rat. J Comp Neurol 246:343–363. https://doi.org/10.1002/cne.902460305

Sosulski DL, Bloom ML, Cutforth T, Axel R, Datta SR (2011) Distinct representations of olfactory information in different cortical centres. Nature 472:213. https://doi.org/10.1038/nature09868. https://www.nature.com/articles/nature09868#supplementary-information

Sternson SM, Atasoy D, Betley JN, Henry FE, Xu S (2016) An emerging technology framework for the neurobiology of appetite. Cell Metab 23:234–253. https://doi.org/10.1016/j.cmet.2015.12.002

Sukikara MH, Mota-Ortiz SR, Baldo MV, Felicio LF, Canteras NS (2010) The periaqueductal gray and its potential role in maternal behavior inhibition in response to predatory threats. Behav Brain Res 209:226–233

Szymusiak R, Satinoff E (1982) Acute thermoregulatory effects of unilateral electrolytic lesions of the medial and lateral preoptic area in rats. Physiol Behav 28:161–170. https://doi.org/10.1016/0031-9384(82)90118-4

Tachikawa KS, Yoshihara Y, Kuroda KO (2013) Behavioral transition from attack to parenting in male mice: a crucial role of the vomeronasal system. J Neurosci 33:5120–5126. https://doi.org/10.1523/jneurosci.2364-12.2013

Tsuneoka Y et al (2013) Functional, anatomical, and neurochemical differentiation of medial preoptic area subregions in relation to maternal behavior in the mouse. J Comp Neurol 521:1633–1663. https://doi.org/10.1002/cne.23251

Tsuneoka Y et al (2015) Distinct preoptic-BST nuclei dissociate paternal and infanticidal behavior in mice. EMBO J 34:2652–2670. https://doi.org/10.15252/embj.201591942

Tsuneoka Y et al (2017) Neurotransmitters and neuropeptides in gonadal steroid receptor-expressing cells in medial preoptic area subregions of the male mouse. Sci Rep 7:9809. https://doi.org/10.1038/s41598-017-10213-4

Watson TC, Cerminara NL, Lumb BM, Apps R (2016) Neural correlates of fear in the periaqueductal gray. J Neurosci 36:12707–12719. https://doi.org/10.1523/jneurosci.1100-16.2016

Weber EM, Olsson IAS (2008) Maternal behaviour in Mus musculus sp.: an ethological review. Appl Anim Behav Sci 114:1–22. https://doi.org/10.1016/j.applanim.2008.06.006

Wei YC et al (2018) Medial preoptic area in mice is capable of mediating sexually dimorphic behaviors regardless of gender. Nat Commun 9:279. https://doi.org/10.1038/s41467-017-02648-0

Wilkinson GS, Miller BAE (2010) Communal nesting among genetically similar house mice. Ethology 77:103–114

Wu Z, Autry AE, Bergan JF, Watabe-Uchida M, Dulac CG (2014) Galanin neurons in the medial preoptic area govern parental behaviour. Nature 509:325–330. https://doi.org/10.1038/nature13307

Yang CF et al (2013) Sexually dimorphic neurons in the ventromedial hypothalamus govern mating in both sexes and aggression in males. Cell 153:896–909. https://doi.org/10.1016/j.cell.2013.04.017

Zha X, Xu X (2015) Dissecting the hypothalamic pathways that underlie innate behaviors. Neurosci Bull 31:629–648. https://doi.org/10.1007/s12264-015-1564-2

Spatial Navigation

7

Wenjun Jin, Han Qin, Kuan Zhang, and Xiaowei Chen

Abstract

The hippocampus is critical for spatial navigation. In this review, we focus on the role of the hippocampus in three basic strategies used for spatial navigation: path integration, stimulus–response association, and map-based navigation. First, the hippocampus is not required for path integration unless the path of path integration is too long and complex. The hippocampus provides mnemonic support when involved in the process of path integration. Second, the hippocampus's involvement in stimulus–response association is dependent on how the strategy is conducted. The hippocampus is not required for the habit form of stimulus–response association. Third, while the hippocampus is fully engaged in map-based navigation, the shared characteristics of place cells, grid cells, head direction cells, and other spatial encoding cells, which are detected in the hippocampus and associated areas, offer a possibility that there is a stand-alone allocentric space perception (or mental representation) of the environment outside and independent of the hippocampus, and the spatially specific firing patterns of these spatial encoding cells are the unfolding of the intermediate stages of the processing of this allocentric spatial information when conveyed into the hippocampus for information storage or retrieval. Furthermore, the presence of all the spatially specific firing patterns in the hippocampus and the related neural circuits during the path integration and map-based navigation support such a notion that in essence, path integration is the same allocentric space perception provided with only idiothetic inputs. Taken together, the hippocampus plays a general mnemonic role in spatial navigation.

Keywords

Hippocampus · Path integration · Space perception · Place cells · Grid cells

7.1 Introduction

Spatial navigation is one of the most fundamental functions necessary for animals to survive. Successful navigation allows animals to find food, water, mates, and breeding grounds and to avoid predators. To effectively navigate from one place to an unseen destination, animals need to choose the proper course or trajectory in accordance with the available information. Animals may take advantage of the information from both proprioceptive and exteroceptive sources using three basic strategies (Fig. 7.1), corresponding to the

W. Jin (✉) · H. Qin · K. Zhang · X. Chen
Brain Research Center and State Key Laboratory of Trauma, Burns, and Combined Injury, Third Military Medical University, Chongqing, China

H. Wang (ed.), *Neural Circuits of Innate Behaviors*, Advances in Experimental Medicine and Biology 1284, https://doi.org/10.1007/978-981-15-7086-5_7

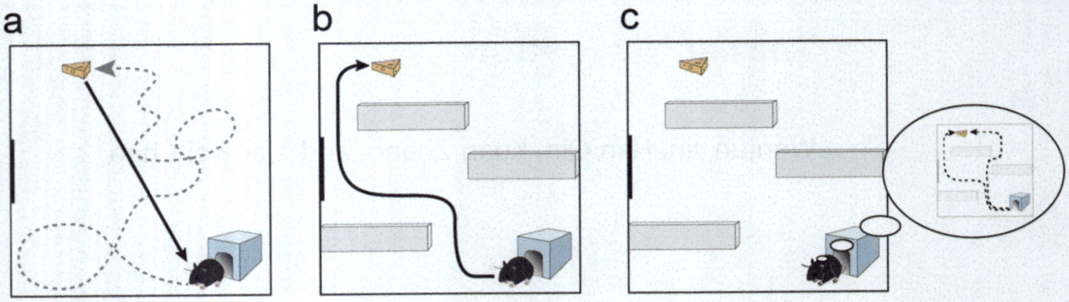

Fig. 7.1 Three navigation strategies. (**a**) Path integration allows the animals to keep track of the vector information relative to the departed nest based on idiothetic inputs. (**b**) Using the stimulus–response strategy, the mouse may reach the reward through several left and right turns on viewing the block and the cue on the wall. (**c**) Map-based navigation allows the mouse to flexibly choose a path to the reward location according to the mental representation (insert) of the environment in the brain

ways in which the space can be represented in different reference frames (Gallistel 1993; O'Keefe and Nadel 1978; Moser et al. 2017). First, animals can rely on internally generated signals to make the movement. It has been shown that information from the vestibular, proprioceptive, and somatosensory systems together with efference copies of motor commands and optic flow signals allows the animal to keep track of its position relative to the point of departure. This type of navigation, which is referred to as path integration, or dead reckoning in marine navigation, allows the animals to navigate in the dark or in situations when other external information is not available or is unnecessary. Second, when information from external sources, such as visual landmarks, sounds, or olfactory or tactile cues, is available, this information could be exploited for spatial navigation in two other ways. In the stimulus–response (S–R) association strategy, which is relevant to the route strategy in the taxon system (O'Keefe and Nadel 1978), a particular cue or landmark is strictly linked with a certain movement for the animals to follow or avoid. It was also found that multiple landmarks can be treated as a snapshot or single integrated cue of the environment to guide the movement (Cartwright and Collett 1982; Collett and Collett 2002). In addition to the S–R strategy where the cue or landmark is individually used, animals have evolved a way to construct a mental representation of the environment from the spatial relationship of the landmarks and choose a navigation course based on the cognitive map of the environment (Tolman 1948). This strategy, which is termed map-based navigation, allows more reliable and flexible navigation to the destination when compared to the path integration and S–R strategy (O'Keefe and Nadel 1978). These three basic strategies can be adapted in various combinations in the real world. For example, an ant may adopt a map-based strategy to look for home only when it identifies that it is in the vicinity of the home (Gallistel 1993).

Previous studies have demonstrated that the hippocampus is essentially involved in map-based spatial navigation (O'Keefe and Nadel 1978; Morris et al. 1982). However, the exact role of the hippocampus in cognitive map-based spatial navigation remains controversial as there are two conflicting opinions (Eichenbaum et al. 2007, 2016; O'Keefe and Nadel 1978; Lisman et al. 2017). In addition, there are inconsistent reports regarding whether the hippocampus is involved in spatial navigation using path integration strategy (Alyan and McNaughton 1999; Maaswinkel et al. 1999; McNaughton et al. 1996; Shrager et al. 2008). In this chapter, we will review the potential role of the hippocampus in spatial navigation under different strategies. More specifically, we focus on the role of the hippocampus in map-based spatial navigation.

7.2 Role of the Hippocampus in Three Navigation Strategies

The hippocampus is a hub receiving information from multiple sensory modalities (Bellistri et al. 2013; Bickford-Wimer et al. 1990; Hartley et al. 2000; Lopez and Blanke 2011; Pereira et al. 2007). More specifically, visual and spatial information typically arrives in the hippocampus through the postrhinal cortex (POR) and then the medial entorhinal cortex (mEC); the other polymodal sensory information, such as auditory and tactile information, reaches the hippocampus through the perirhinal cortex (PER) and lateral entorhinal cortex (lEC) pathway (Witter and Amaral 2004). Besides the input from the afferent cortical structures, the hippocampus receives heavy subcortical input, such as that from the thalamus, via the PER and POR. The vast amount of information input into the hippocampus is consistent with the important role of the hippocampus in spatial navigation.

7.2.1 Role of the Hippocampus in Path Integration

Path integration has been found in a broad range of animals, including insects, birds, rodents, and humans (Etienne and Jeffery 2004; Gallistel 1993; Heinze et al. 2018; Mittelstaedt and Mittelstaedt 1982, 1980). Considering the vast anatomical differences between these species, we will discuss path integration only in mammals.

Currently, whether the hippocampus is involved in path integration in mammals is controversial. Patients with hippocampal or entorhinal lesions are able to point out the direction to a start location and estimate the distance as accurately as controls while blindfolded (Shrager et al. 2008), suggesting that the hippocampus is not required for path integration in humans. Although path integration has been reported to be impaired in some animals after fimbria-fornix (FF), mEC, or direct hippocampal lesions (Kim et al. 2013; McNaughton et al. 1996; Parron and Save 2004; Whishaw et al. 2001; Whishaw and Maaswinkel 1998), the overall conclusions from these experiments in animals remain mixed (Alyan and McNaughton 1999; Etienne and Jeffery 2004; Poucet and Benhamou 1997). A close inspection of these results shows that the typical kinematic profiles of the impaired path integration in animals with vestibular lesions are very different from the ones recorded in hippocampal-lesioned animals in the dark (Fig. 7.2). Animals with vestibular lesions usually run to the correct destination through an indirect route, yielding a markedly longer path (Wallace et al. 2002). In contrast, animals with hippocampal lesions often take a direct route to the wrong destination (Parron and Save 2004; Whishaw and Maaswinkel 1998). Notably, the movement of vestibular-lesioned animals is quickly restored to the direct route once the light is turned on (Wallace et al. 2002). Therefore, the path integration in the animals with hippocampal lesions may have been intact, and the reported impairment in path integration tasks may be simply because the animals lost the memory of the refuge to return to. Actually, the involvement of the hippocampus in path integration has been uncertain from the beginning, and instead, working spatial memory may take part in path integration (Whishaw and Maaswinkel 1998).

The distinct results from the path integration tests in animals and humans may be due to the differential capacity of spatial working memory between humans and rats, as revealed by a study with a classic path integration task in the Squire lab (Kim et al. 2013). Patients with hippocampal lesions carried out the path integration task as well as the controls did when the outward path was relatively direct and the task could be finished within 20 s. This time range is consistent with the failure time observed in H.M. in the delayed paired comparison test (O'Keefe and Nadel 1978). On the other hand, it was found that the performance of path integration deteriorated quickly in normal rats once the travel distance was longer than 2 m, more than one turn or a time range longer than 6 s was used (Kim et al. 2013). Therefore, the hippocampus seems to not be required for the performance of path integration until the demands on spatial working

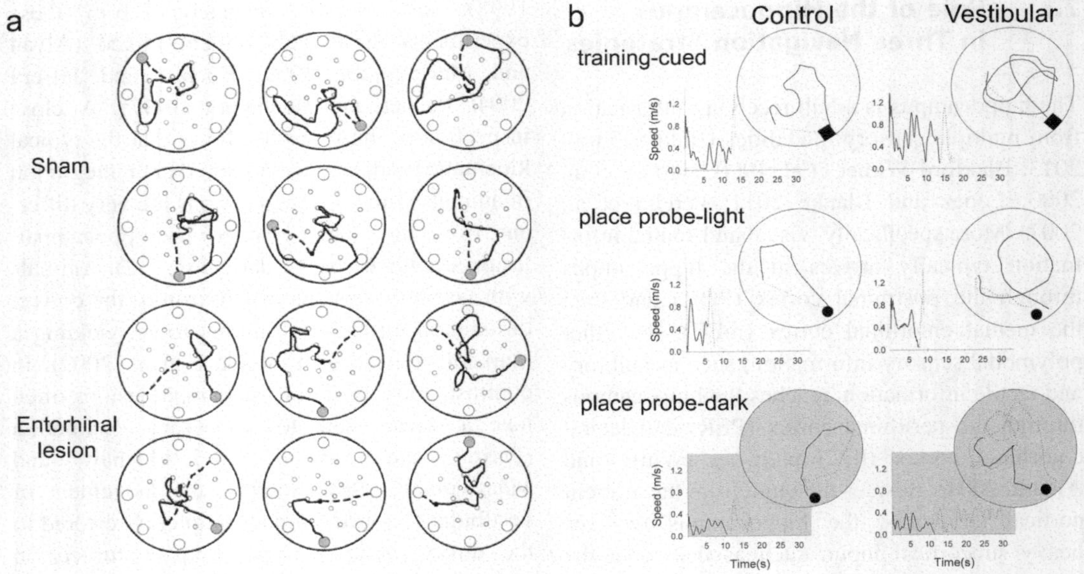

Fig. 7.2 Comparison of animal movements with entorhinal and vestibular lesions in the dark. (**a**) Examples of typical kinematic profiles from animals with sham and entorhinal lesions (Parron and Save 2004). Rats leave a refuge from the starting holes (gray circle) to find the food pellet hidden in the reward cup (black circle) and carry it back to the refuge. The outward and return paths are plotted with full and dotted lines, respectively. In both the sham and entorhinal lesion groups, the food pellet is either located in the central cup (top row) or randomly chosen cup (second row). Rats in the sham group always directly return to the refuge. Entorhinal-lesioned rats typically randomly choose a destination hole and run directly to it. (**b**) Movements of animals with vestibular lesions and control animals (Wallace et al. 2002). Similar tasks as in (**a**) are undertaken, and rats need to find the food pellet and carry it back to the refuge. The outward paths are shown as solid lines, and the return paths are shown as dotted lines. The top, middle, and bottom rows are the representative kinematic profiles of the control and vestibular-lesioned animals during cued training, probe testing with normal vision, and probe testing in the dark, respectively. The black rectangle indicates that there is a black box positioned over the home base and is clearly visible to the rats on the table. The black box is absent during the probe tests, and the home base is marked with a black dot. While all rats can find the home base correctly, the path of rats with vestibular lesions is typically indirect in the dark; therefore, animals take much more time and a longer path to return

memory are high with large information loading (De Nigris et al. 2013; Jones and Wilson 2005; Leszczynski 2011; Spellman et al. 2015; Yoon et al. 2008). Thus, it is not surprising to observe the recruitment of the hippocampus in navigation with path integration in humans (Suthana et al. 2012).

7.2.2 Role of the Hippocampus in the S–R Strategy

It has been suggested that the hippocampus is involved only during the early phase of the S–R strategy practice and that the caudate nucleus is required when the pairing of the stimulus and response becomes automatic and turns into a form of procedural memory or habit after extensive training (McDonald and White 1993; Packard and Knowlton 2002; Packard and McGaugh 1996; Seger and Spiering 2011). Greater activation of the caudate nucleus was observed when following a well-learned route, which is in contrast with the greater hippocampal activation during searching for a new route in the same individuals (Hartley et al. 2003). Similarly, distinct involvement of the hippocampus and striatum was also observed during context memory and reinforcement learning (stimulus–response associations) in attention tests (Goldfarb et al.

2016). Consistently, lesions of the hippocampus cause impairments in the Morris water maze, and lesions of the dorsal striatum result in defects in the Morris water maze with a visible platform (Devan et al. 1999; McDonald and White 1994; Morris et al. 1982). However, while the hippocampus was not required for the animals to make left and right turns in the continuous T-maze spatial alternation task, hippocampal-lesioned animals failed the behavior task once a 2- or 10-s delay was imposed between the alternations (Ainge et al. 2007a). Therefore, the involvement of the hippocampus in the S–R strategy is dependent on how this strategy is conducted. The hippocampus is not required in the S–R strategy only when it becomes a habit or procedural memory.

7.2.3 Role of the Hippocampus in Map-Based Navigation

Currently, there are two different opinions regarding the role of the hippocampus in map-based spatial navigation (Buzsáki and Moser 2013; Eichenbaum 2000; Eichenbaum et al. 2016; Eichenbaum and Cohen 2014; O'Keefe and Nadel 1978; O'Keefe 1999) system. The cognitive map theory, which was established with the discovery of place cells (O'Keefe and Nadel 1978; O'Keefe and Dostrovsky 1971), holds that the hippocampus constructs a spatial map of the environment and shares it with the rest of the brain (Eichenbaum et al. 2016; O'Keefe and Nadel 1978; O'Keefe 1999). The cognitive map theory was further strengthened with the subsequent finding of head direction cells, grid cells, border cells, and other spatially specific cells in the medial entorhinal cortex and other related structures (Hafting et al. 2005; Kropff et al. 2015; Lever et al. 2009; Ranck 1984; Solstad et al. 2008; Taube et al. 1990a; Taube 1995). Besides the view that the map-based spatial navigation is the solo or primary function of the hippocampus, the alternative view is that the hippocampus plays a general role in memory function, and the hippocampus is thought to support not only maps of physical space but also a general map of cognition covering both spatial and nonspatial relationships (Cohen and Eichenbaum 1993; Lisman et al. 2017; Schiller et al. 2015). This view is also supported by many experiment data, such as the hippocampal neurons encoding both the spatial and nonspatial cognitive signals (Aronov et al. 2017). However, the systematic comparison of all the spatial-specific firing in the hippocampus and related limbic circuits shed new light on the source of these spatially specific firing pattern and therefore brings about new consideration about the role of the hippocampus in map-based spatial navigation.

7.3 Spatially Specific Activity in the Hippocampus and the Related Limbic Circuits

Since the discovery of place cells in the hippocampus, additional spatial encoding cells were subsequently found in the entorhinal cortex and other limbic circuits (Goodridge and Taube 1997; Hafting et al. 2005; Kropff et al. 2015; Lever et al. 2009; Sargolini et al. 2006; Solstad et al. 2008). In particular, head direction cells with preferred directional discharges were recorded in the postsubiculum (PoS) (Ranck 1984) and then in other cortical and subcortical structures (Blair et al. 1998; Chen et al. 1994; Cho and Sharp 2001; Sargolini et al. 2006; Taube et al. 1990a; Taube 1995). In the mEC, grid cells with a regular triangular spatial pattern (Fyhn et al. 2004; Hafting et al. 2005), boundary cells with specialized discharges along the border (Savelli et al. 2008; Solstad et al. 2008) together with cells encoding running speed (Kropff et al. 2015) and carrying other conjunctive spatial firing patterns were detected successively (Grieves and Jeffery 2017). These distinct spatial representations are distributed in a broad range of structures with a bunch of cells typically clustered in the mEC. Next, we will focus on the firing properties of place cells, grid cells, and head direction cells (Fig. 7.3). These cell types were chosen because they are the most frequently studied and are distributed in different anatomical structures.

It is worth noting that all these types of spatially specific firing patterns are allocentric, that

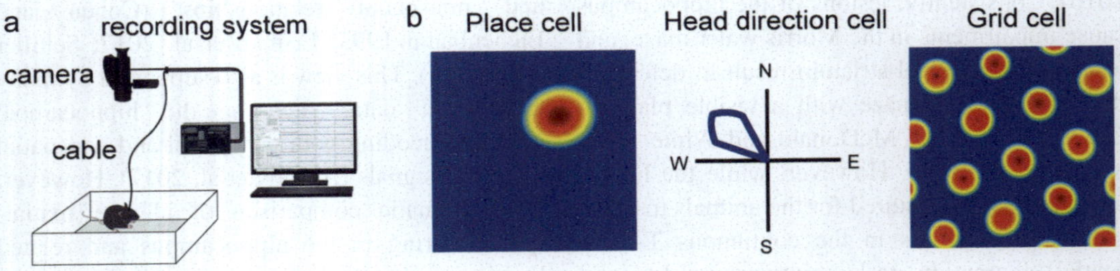

Fig. 7.3 Place cells, head direction cells, and grid cells. (**a**) A typical experimental setup to measure the spatially specific activity of cells from freely behaving rats chronically implanted with extracellular electrodes. (**b**) The activity of the place cell and grid cell is typically illustrated as a heat plot of the firing rate against the animal location. Here, the place field is in the northeast part of the enclosure, and the discharge of grid cells is occurred in multiple small regions distributed throughout the environment in a regular hexagonal pattern. The middle polar plot, presenting the firing rate as a function of head direction, shows that this head direction cell has a preferred firing in the northwest direction

is, constructed based on external references. The advantage of allocentric spatial encoding is that the represented allocentric spatial information is stable and will not be changed when the animal moves (Fig. 7.4). Theoretically, the constancy of the allocentric spatial relationship may allow its offline utilization, as long as the information is saved and retrieved properly. The allocentric spatial representation is believed to be converted from the egocentric perspective (Gallistel 1993) although the exact mechanism remains elusive. Egocentric spatial encoding, which is self-referential, has been detected in the parietal cortex and associated regions but not in the hippocampus (Snyder et al. 1998; Stein 1989; Wilber et al. 2014). In contrast to allocentric spatial firing, egocentric spatial representations change according to the animals' standing location and orientation and therefore best fit the purpose of online information processing.

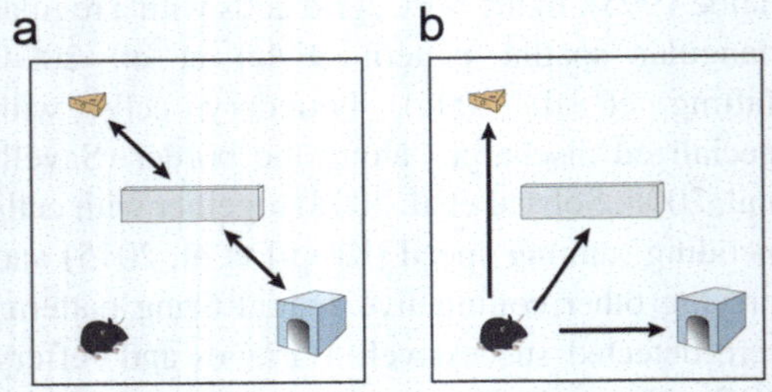

Fig. 7.4 Allocentric versus egocentric spatial encoding. Within the allocentric spatial representation, the location of cheese is defined relative to the location of other objects, including the block, and perhaps also the corner of the enclosure (**a**). The egocentric location of cheese is defined relative to the body axis of the observer (self) as left or right, front or back (**b**). Thus, the egocentric location of cheese is correct only when the animals stay at that position. As a comparison, the allocentric location is constant regardless of where the animal is, making it suitable for offline planning

7.3.1 Place Cells

Place cells are hippocampal pyramidal cells that fire specifically when the animal traverses a certain region of the environment (Alme et al. 2014; Leutgeb et al. 2004; O'Keefe and Dostrovsky 1971). The area of high firing rate, which is defined as the cell's place field, usually takes up a contiguous and irregular field (O'Keefe 1979; O'Keefe and Conway 1978; Wilson and McNaughton 1993). Although the discharge of place cells can also be affected by direction or other nonspatial factors (Ainge et al. 2007b; Gothard et al. 1996; Lee et al. 2006; Moser et al. 2017; Muller et al. 1994; Redish et al. 2000; Wood et al. 2000), the place field is determined and influenced primarily by the salient cues or landmarks within the environment space (Grieves and Jeffery 2017; O'Keefe and Nadel 1978; Poucet et al. 2000). For example, place cells adapt their specific firing locations strictly following the rotation of a cue card in a cylinder with a uniform interior surface (Muller and Kubie

1987). On the other hand, chronic recording up to 6 months shows that the place field is typically retained within the same environment (Muller et al. 1987; Thompson and Best 1990). Experiment with multiple distal cues inside the environment further shows that the place field can resist partial removal of the salient cues, suggesting that the location of the place field is not determined by individual cues, but rather by the overall configuration of the cues (O'Keefe and Conway 1978).

Place cells change their location-specific firing once the surrounding environment is changed (Alme et al. 2014; Bostock et al. 1991; Colgin et al. 2008; Kentros et al. 1998; Leutgeb et al. 2005; Markus et al. 1995). This phenomenon, which is termed remapping, takes two essentially independent forms as rate remapping and global remapping (Leutgeb et al. 2005). Rate remapping occurs when only the nonspatial features of the environment, such as the wall color of the enclosure, are changed. The firing rate, but not the place field of the place cells, is changed during the rate remapping. When animals are placed in different containing boxes or moved into different rooms, global remapping occurs with the change of both the place field and the firing rate.

One unique feature of global remapping is the unpredictable change in the place field. There is no correlation between the place field in one environment and the place field in another environment (Muller and Kubie 1987; Wilson and McNaughton 1993). Therefore, many place cells are silent in some environments and present place fields in other environments (Muller and Kubie 1987; O'Keefe and Speakman 1987; Thompson and Best 1989). Such random remapping is consistent with the irregular distribution pattern of the place fields as there is no consistent topographical relation between the place fields and the anatomical location of the corresponding place cells (Dombeck et al. 2010; Leutgeb et al. 2004; Mizuseki et al. 2012; O'Keefe 1976; O'Keefe et al. 1998; Wilson and McNaughton 1993). Such a random pattern of the place fields is fit for the representation of distinct spatial environments by the place cell population. This phenomenon has been verified in a recording of place cells in 11 different rooms with minimal overlap (Alme et al. 2014).

While visual information takes precedence in determining the firing field, visual cues is not required for place cells to maintain their unique spatially specific firing patterns. It was found that place fields are retained in darkness if the animals remain in the box before the lights are turned off (Markus et al. 1994; Quirk et al. 1990; Zhang et al. 2014). In contrast, place fields are disrupted if the rats are put into the maze in the dark (Mcnaughton et al. 1989). It has been suggested that stable spatially selective firing when visual information is absent is maintained through path integration with idiothetic information from local olfactory or tactile cues combined with motion-related cues (Poucet et al. 2000). In another relevant circumstance with masked external cues, rotating the animal gently outside the enclosure before putting it back in the recording box almost always led to a corresponding rotation of the place field, while rotating the recording box alone seldom triggered field rotation (Jeffery et al. 1997). What makes the situation more complicated is that place cells exist in blind animals, suggesting that the spatial representation could also be supported by information from other sensory modalities (Hill and Best 1981; Save et al. 1998). More interesting is that the place cells in blind rats discharge only after the animals make physical contact with the object after the animals walk into the place field (Save et al. 2000).

The phenomenon of place coding is relatively common, as place cells or place-like "spatial view" neurons have been identified in rats and mice, bats, monkeys, and humans (Ekstrom et al. 2003; Kentros et al. 2004; Matsumura et al. 1999; Rolls 1999; Rolls et al. 1997; Ulanovsky and Moss 2007). It is worth noting that the place-like "spatial view" cells observed in the primate hippocampus behave differently from those in the hippocampus of rodents as they are activated when the animal looks at a particular region of the environment, independent of the animal's physical location (Rolls 1999; Rolls et al. 1997).

7.3.2 Grid Cells

Grid cells are a set of pyramidal cells and satellite cells in the mEC and the pre- and parasubiculum which present firing in a hexagonal grid pattern spanning the whole space of the environment (Boccara et al. 2010; Fyhn et al. 2004; Hafting et al. 2005; Sargolini et al. 2006). This unique grid pattern is characterized by scale (distance between firing fields), orientation (angle between grid axes and environment), and phase (relative position of firing peaks) (Fyhn et al. 2004; Hafting et al. 2005; Stensola et al. 2012). It was revealed that the grid cells are regularly distributed as grid cells nearby present similar scale and orientation (Hafting et al. 2005; Heys et al. 2014), and there is a trend toward increased scale for the cells distributed along the dorsal-ventral axis of the mEC (Brun et al. 2008b; Stensola et al. 2012). With all these unique spatial properties, grid cells have been suggested to be important for estimating the distance traveled during navigation (Moser and Moser 2008).

Previous studies have identified factors affecting the grid firing pattern. First, the grid firing pattern is primarily determined by the environmental geometry (Krupic et al. 2015). When the landmarks of the environment are rotated, the grid orientation and phase follow the rotation of the landmark to the same degree, while the grid scale remains unchanged (Hafting et al. 2005). Modification of the geometry of the environment often results in firing pattern changes (Barry et al. 2007; Stensola et al. 2012). Local changes to the configuration of the environment induce a shift in the grid fields near the changed wall (Krupic et al. 2015, 2016, 2018; Stensola et al. 2015). Second, grid firing has also been suggested to be controlled by internal movement signals. When rats walk into the interconnected arms of a modified hairpin maze, the firing locations in the normal and shortcut-modified arms with the same direction are equally determined by the walked distance (Derdikman et al. 2009). When rats are placed into the box before the lights are turned off, the firing patterns of grid cells are retained under complete darkness (Fyhn et al. 2007; Hafting et al. 2005). Although there have been mixed results showing that the firing pattern of grid cells is severely disrupted in darkness, the conflicting results may be due to the interspecies difference as all the results have been measured in mice (Chen et al. 2016; Pérez-Escobar et al. 2016).

The grid firing pattern seems to be a common phenomenon in the central nervous system of mammals. Since the initial discovery of grid cells in rats, grid or grid-like firing patterns have been detected in other rodents (Fyhn et al. 2008; Hafting et al. 2005), bats (Yartsev et al. 2011), monkeys (Killian et al. 2012; Killian and Buffalo 2018), and humans (Doeller et al. 2010; Jacobs et al. 2013; Julian et al. 2018; Nau et al. 2018). More specifically, the grid firing pattern discovered in rodents and bats is directly related to the physical space, but the grid-like pattern detected in the entorhinal cortex (EC) of monkeys (Killian et al. 2012; Killian and Buffalo 2018) and humans (Julian et al. 2018; Nau et al. 2018) covers the visual space they are watching. Even more intriguing, grid-like patterns of activation have also been recorded in the mEC when subjects imagine spatial content or engage in certain forms of conceptual thinking (Bellmund et al. 2016; Constantinescu et al. 2016; Horner et al. 2016).

7.3.3 Head Direction Cells

Head direction (HD) cells are a set of neurons first detected in the PoS of the rat that maximally fire when the animal's head is oriented to a particular "preferred firing direction" in the horizontal plane (Ranck 1984; Taube et al. 1990a, 1990b). The preferred firing direction is relative to the spatial configuration of the environment as the rotation of the salient visual landmarks in the environment leads to a corresponding shift in the preferred firing direction of HD cells (Knierim et al. 1995). More specifically, HD cells bind more strongly to distal landmarks when there is conflict between the rotations of the proximal and distal cues (Yoganarasimha et al. 2006; Zugaro et al. 2001). When visual cues are placed in conflict with idiothetic cues, the spatial information derived from visual landmarks usually overrides that of idiothetic inputs (Blair and Sharp 1996; Goodridge and Taube 1995). When there is only idiothetic information available, such as situations with the lights off or the animals being blindfolded, directional firing can be retained as long as the HD cells have previously established a directional preference (Goodridge et al. 1998; Taube et al. 1990b) although small drifts in the preferred firing direction can be observed over long periods of time (Goodridge et al. 1998; Knierim et al. 1995). The preferred firing direction in the dark has been suggested to be retained by the action of vestibular cues through path integration (Blair and Sharp 1996).

Since the initial discovery of HD cells by Ranck and his colleagues in the PoS, HD cells have been recorded in multiple cortical and subcortical regions, such as the anterodorsal nucleus (ADN) of the anterior thalamus (Taube 1995), lateral mammillary nuclei (LMN) (Blair et al. 1998; Stackman and Taube 1998), lateral dorsal thalamic nucleus (LDN) (Mizumori and Williams 1993), retrosplenial cortex (RSC) (Chen et al. 1994; Cho and Sharp 2001), dorsal striatum (Wiener 1993), posterior cortex (Chen et al. 1994), medial precentral cortex (PrCM) in the frontal lobe (Mehlman et al. 2018), and mEC (Sargolini et al. 2006). The tuning curves of HD cells across different brain areas are remarkably similar (Taube and Bassett 2003). As these HD-containing regions generally have strong anatomical interconnections, there seems to be a hierarchical processing of head direction information from the vestibular nuclei to the LMN, ADN, and PoS (Dumont and Taube 2015).

The hierarchical structure of the HD system is supported by the lesion experiments. Lesions of the ADN disrupt HD cell activity in the PoS (Goodridge and Taube 1997), while lesions of the PoS (Goodridge and Taube 1997), LDN (Mizumori and Williams 1993), or posterior parietal cortex leave the HD cells intact in the ADN (Calton et al. 2008). Bilateral lesions of the LMN or dorsal tegmental nucleus (DTN) of Gudden disrupt HD cell firing in the ADN and PoS (Blair et al. 1998). Besides the hierarchical processing, it seems that all the HD cells have shared input as simultaneous recording in more than one HD cell reveals that their preferred firing directions always change in register with different environmental contexts (Taube et al. 1990b).

7.3.4 The Relationship of Spatial Encoding by Place Cells, Grid Cells, and HD Cells

It is known that the location, distance, and direction information are dissociable in a conventional map (Wood et al. 2000). How are the location, distance, and direction representations different in the nervous system? We tried to address this question by examining the potential interactions between place cells, grid cells, and HD cells. Here, HD cells in the LMN, ADN, and PoS circuits were considered since HD cells in these structures are the most basic components of the HD system (Brown et al. 2002).

7.3.4.1 Place Cells Versus Grid Cells

It is taken for granted initially that the firing of place cells is influenced by grid cells (Brun et al. 2002). First, it has been verified that hippocampal place cells receive inputs directly from grid cells, together with border cells, HD cells, and some other neurons in the mEC (Zhang et al. 2013). In addition to direct projections from mEC layer II neurons into the dentate gyrus (DG) and CA3, there are projections from neurons in layer III of mEC (MECIII) to the proximal part of CA1 (Henriksen et al. 2010; Witter and Amaral 2004). Second, simultaneous recording in the hippocampus and mEC during environment transformations has confirmed that the remapping in the place cells occurs along with changes in grid firing patterns in the mEC (Fyhn et al. 2007; Stensola et al. 2012). Third, bilateral lesions of the mEC result in unstable place fields (Brun et al. 2008a; Hales et al. 2014; Miller and Best 1980; Schlesiger et al. 2015; Van Cauter et al. 2008), and hyperpolarization of the superficial mEC neurons leads to hippocampal remapping (Zhao et al. 2016). Accompanying the unstable firing or artificial remapping caused by mEC manipulation, the spatial memories in hippocampus-dependent tasks are also impaired (Hales et al. 2014; Kanter et al. 2017; Parron and Save 2004; Steffenach et al. 2005).

However, subsequent experiments reject such simplicity with controversial results. In one approach, Kanter et al. developed a chemogenetic method to depolarize or hyperpolarize almost exclusively stellate cells in mEC layer II (MECII) without modification of the grid firing pattern (Kanter et al. 2017). They found that the depolarization, but not hyperpolarization, of the MECII stellate cells caused place cell remapping. When the theta rhythm is reduced with local fusion of lidocaine in the septum, the grid firing pattern in the mEC is erased while the firing of place cells, together with some other non-grid spatial cells in the mEC, is largely retained (Koenig et al. 2011). In another relevant study, hippocampal global remapping occurs in a novel environment when grid cell firing is disrupted through the inactivation of the medial septum by muscimol (Brandon et al. 2014). Remapping of the hippocampal place field also occurs after direct inactivation of mEC neurons with pharmacologic and optogenetic methods (Miao et al. 2015; Ormond and McNaughton 2015; Rueckemann et al. 2016). A recent experiment with a complete lesion of the bilateral mEC superficial layers demonstrated that global remapping occurred in a novel environment without any input from the mEC (Schlesiger et al. 2018).

On the other hand, while there are only indirect projections from the CA1 pyramidal cells to the superficial layers of mEC via the subiculum (Witter 1993), place cells seem to have a strong influence over the grid cells. Bilateral hippocampal lesions with ibotenate led to grid firing with decreased spatial coherence and increased distance between neighboring grid fields (Fyhn et al. 2004). When hippocampal activity was inhibited through the local infusion of the GABAA receptor agonist muscimol, the spatial periodic grid pattern of the grid cells was erased and turned into directional tuning (Bonnevie et al. 2013).

7.3.4.2 HD Cells Versus Place Cells and Grid Cells

HD cells seem to fire independent of grid cells and place cells. First, grid cells have little influence on HD cells as mEC lesions result in almost intact direction-specific firing in the ADN HD cells (Clark and Taube 2011). The control of HD firing by salient visual cues is also retained in mEC-lesioned animals. Second, HD cell signals were generated and properly maintained across days when hippocampus-lesioned animals were put into a novel environment (Golob and Taube 1997).

Instead, HD cells are differentially involved in the proper firing of place cells and grid cells. The firing of place cells is mainly intact in LMN- or ADN-lesioned animals (Calton et al. 2003; Sharp and Koester 2008), suggesting that the head direction information represented by HD cells is not required for place cell activity. However, lesions of the PoS, which is downstream from the ADN, result in unstable place field representations when the visual cue is removed. Place fields also present random shifts following cue card rotations in PoS lesioned animals (Calton et al. 2003). Meanwhile, the HD signal is required for grid cell activity as the grid firing pattern is abolished after ADN inactivation or lesion (Winter et al. 2015a, b). The grid pattern is further suggested to be conveyed from the mixture of the place information together with head direction and border information (Krupic et al. 2016). Consistently, when rats are passively transported in a clear plastic cart, grid-specific firing patterns are abolished while the HD cell firing is spared (Winter et al. 2015a, b).

The interaction between HD cells, place cells, and grid cells is consistent with their distinct developmental timing. HD cells develop earliest as direction tuning is present before eye opening around postnatal day 11 (P11), and mature-like firing can be detected around P16-18 (Bjerknes et al. 2015; Langston et al. 2010). Place cells are present around P16-18, and the grid cells develop slowest and appear around P18-28 (Bjerknes et al. 2018; Langston et al. 2010; Wills et al. 2010).

Interestingly, border cells develop at the same pace as HD cells (Bjerknes et al. 2014).

7.3.5 Shared Properties of Place Cells, Grid Cells, and HD Cells

In addition to the distinct interactions among place cells, grid cells, and HD cells, the spatially specific firing pattern of these cells has many common attributes. First, as mentioned earlier, salient visual landmarks in the environment have strong control over the firing of place cells, grid cells, and HD cells. Simultaneous recording across multiple neural circuits has revealed that these spatially specific firing patterns change in register following the rotation of the external cue (Hafting et al. 2005; Knierim et al. 1995; Muller and Kubie 1987). Second, all these different patterns of spatial firing can be maintained in the dark (Fyhn et al. 2007; Goodridge et al. 1998; Hafting et al. 2005; Markus et al. 1994; Quirk et al. 1990; Taube et al. 1990b; Zhang et al. 2014). However, two recent reports show that the grid firing patterns in mice are disrupted in darkness and only partially rescued after several trials of training (Chen et al. 2016; Pérez-Escobar et al. 2016). This disparity may be due to inter-species differences since the results from place cells, grid cells, and HD cells in rats are all consistent. Third, when the vestibular system is lesioned or inactivated, the directional firing in the HD system, together with the hippocampal place-specific firing and unique grid firing pattern in the mEC, is all erased (Jacob et al. 2014; Russell et al. 2003; Stackman et al. 2002; Stackman and Taube 1997).

Notably, all these spatially specific activities are influenced by animal constraints. When rats are restrained tightly by hand, discharge of the HD cells is almost abolished during passive motion (Knierim et al. 1995; Taube 1995; Taube et al. 1990b). In this case, the proprioceptive input is unlikely the determining factors since the directional firing is retained but at a reduced rate when the animals are relatively loosely restrained. The activities of the place cells are also impaired when rats are transported from one location to another while being restrained by wrapping the body and limbs in a towel fastened with clips (Foster et al. 1989). However, the firing of place cells, grid cells, and HD cells is roughly normal in head-restrained animals during virtual navigation (Chen et al. 2018). It seems that it is not the head restraint or proprioceptive input but rather the external pressure or interference that leads to impairments in the spatially specific firing.

Taken together, the existence of these shared properties of place cells, grid cells, and HD cells implies that these spatially specific firing patterns may have shared mechanisms and are coherent parts of an interconnected system. Within this system, the tuning curves of HD cells, place fields of place cells, and the firing vertices of grid cells are tightly coupled following the cue card rotation (Fyhn et al. 2007; Knierim et al. 1995; Stensola et al. 2012; Yoganarasimha and Knierim 2005). Knierim et al. showed that HD cells and place cells always maintain the same bearing relative to each other, regardless of whether these spatial firing patterns follow the visual cue or drift away during disorientation sessions (Knierim et al. 1995). In addition, the coherent firing of border cells together with grid cells and HD cells in the mEC further implies that border cells are also part of this shared system (Solstad et al. 2008). Considering the roughly independent firing of place cells and HD cells (here, the firing of grid cells is not considered since it mainly depends on both place information from the hippocampus, direction information from the HD system, and perhaps border information in the mEC), the shared mechanism behind the spatially specific firing of place cells, grid cells, and HD cells is unlikely to be located in the hippocampus, EC, or the HD circuits.

One cautionary note is that there is a contradictory result reporting that the activities of place and grid cells in mice are not necessarily coherent (Chen et al. 2019).

7.4 Allocentric Space Perception Is Behind All the Spatially Specific Firing Patterns

The term allocentric space perception is derived from absolute, or non-egocentric space perception and can be traced back to the cognitive map idea initially raised by Edward C. Tolman (O'Keefe and Conway 1978). Tolman proposed that the S–R theory could not provide the proper explanation about the way-finding problem in a complex maze design and that there should exist a mental representation about the environment in the rat brain that "*in the course of learning, something like a field map of the environment gets established in the rat's brain… And it is this tentative map, indicating routes and paths and environmental relationships, which finally determines what responses, if any, the animal will finally release*" (Tolman 1948). The thereafter discovery of place cells in the hippocampus CA1 promoted this idea further, and the hippocampus was hypothesized to be the neural substrate of this mental representation of the environment (O'Keefe and Nadel 1978; O'Keefe and Dostrovsky 1971). However, the many common characteristics of place cells, HD cells, grid cells, and border cells, which are located in a broad range of anatomical circuits, are in favor of the argument that this mental representation of the environment, which has also been termed absolute, or non-egocentric space perception (O'Keefe and Nadel 1978; Tolman 1948), may be the final driving force behind all these allocentric spatial firing patterns or at least provides the allocentric components.

Contrary to the classic cognitive map theory that the awareness of an animal's location in space depends on the activity of place cells in the hippocampus, a further speculation is that the hippocampus receives and processes the content of allocentric space perception for the purpose of information storage or retrieval. Thus, all the spatially related firing patterns in place cells, grid cells, etc., are actually the unfolding of these intermediate stages during the progressive spatial information processing of the allocentric space perception throughout the hippocampus-entorhinal circuits (Behrens et al. 2018). Consistent with this speculation, a correlation analysis of a large volume of simultaneously recorded grid cells revealed that the correlation within the same module was most likely derived from common input (Tocker et al. 2015). Allocentric spatial information may be simultaneously processed with other nonspatial or egocentric spatial information (Lisman et al. 2017; Meshulam et al. 2017; Wood et al. 2000; Young et al. 1994). While the significance of such complex information processing that involves so many regions, including the RSC, PoS, POR, PER, mEC, CA1, CA3, and DG, remains unclear, damage to any of these intermediate structures will result in the impairment of spatial memory (Alvarado and Bachevalier 2005; Epstein 2008).

It is now understood that allocentric space perception and spatial information storage are undertaken by two different anatomical domains. The dissociation between these two functions can be observed under certain clinical circumstances with hippocampal resection (Maguire et al. 2006; Rosenbaum et al. 2000; Teng and Squire 1999). As reported, the amnesic patient (E.P.) could navigate successfully from his old homes to different locations in a familiar area or make alternative routes during navigation within the familiar area, but could not do almost any navigation in his current (new) environment. While he had normal allocentric space perception, he could not save the content of allocentric space perception via the hippocampus for offline usage. In another test with minimized demand on information saving through the hippocampus, patients presented proper navigation by following the path instruction, which was printed and readable on a hand-held map (Urgolites et al. 2016). Thus, in this scenario, the declarative memory view and the spatial navigation view of hippocampal function are reconciled (Eichenbaum 2017; Squire and Zola 1996).

As perception is susceptible to prior experience, this is also true regarding the discharge of place cells, grid cells, and HD cells. For example, when animals are trained in a fixed sequence in an environment with two identical compartments arranged in parallel, and the animals are tested once to start in a reversed order, a firing pattern similar to that in the default starting box is detected during the first visit and reverts to the familiar pattern thereafter (Skaggs and McNaughton 1998). Sometimes, the allocentric space perception may deviate far from the actual environment. Muller et al. changed the position of a visual cue on the wall in the presence of the rats and found that the place field was completely remapped when the cue had returned to its original position after an initial 180° rotation followed by four 45° rotations (Alexander and Muller 1997) (Fig. 7.5). In another experiment, O'Keefe and Speakman showed that the place firing pattern can be aligned only with the rat's choice of the goal arm, not the designated arm (O'Keefe and Speakman 1987), suggesting that the place-specific firing pattern reflects only what the animal gets.

7.4.1 Neural Substrate of the Allocentric Space Perception

The neural substrate underlying allocentric space perception remains to be identified. One potential candidate is the retrosplenial cortex (RSC) (Clark et al. 2018; Mitchell et al. 2018). The RSC projects to the EC and is interconnected with the posterior parietal cortex (PPC) together with the subicular complex (Witter 1993). The RSC is also suggested to relay visual information directly from the visual areas into the mEC via the PoS (Kononenko and Witter 2012; van Groen and Wyss 1992; Vogt and Miller 1983). Thus, its anatomical location within the limbic system makes it suitable to convert egocentric information into an allocentric format (Chen et al. 1994; Vann and Aggleton 2005). In addition to this transform, the RSC has also been suggested to transform allocentric representations into egocentric representations (Vann et al. 2009). In the human RSC, activation was observed when a conversion from an egocentric viewpoint into an allocentric reference frame was required (Lambrey et al. 2012; Vass and Epstein 2013). Defects in direction sensation were observed on patients with right retrosplenial lesion (Takahashi et al. 1997).

However, there are other conflicting results about the RSC in spatial navigation (Mitchell et al. 2018). Proper HD firing, together with the intact control of the preferred firing direction with the landmark, has been reported in RSC-lesioned rats (Golob and Taube 1999). Inactivation of the retrosplenial cortex with lidocaine results in impaired place cell remapping, but only in dark conditions (Cooper and Mizumori 1999, 2001). Clark et al. proposed that the transformation of information from the egocentric to the allocentric format is not completely conducted in the RSC but rather in both the RSC and PPC, with different task divisions between them (Clark et al. 2018).

Another structure of interest is the claustrum. The claustrum is listed as a candidate because cells encoding place, boundary, and object information are detected in the anterior claustrum of the freely moving rats (Jankowski and O'Mara 2015). Spatial tasks are reported to be impaired after claustrum lesions or optogenetic silencing (Grasby and Talk 2013; Kitanishi and Matsuo 2017). Thus, as the claustrum receives major input from all the neocortical areas and subcortical areas, including the thalamus and hypothalamus (John et al. 2014), either the claustrum itself or one of its upstream regions may be the source of the allocentric spatial signal. It is intriguing to note that the claustrum has extensive direct and indirect connections with the hippocampal circuits (Park et al. 2012; Witter et al. 1988), and there is limited afferent input from the RSC into the claustrum (Zingg et al. 2018).

7.4.2 Path Integration Is the Allocentric Space Perception in Mammals Provided with Idiothetic Inputs

Path integration has been demonstrated to be involved in the firing of place cells, grid cells, and HD cells (Fyhn et al. 2007; Hafting et al. 2005; Markus et al. 1994; O'Keefe 1976; Quirk

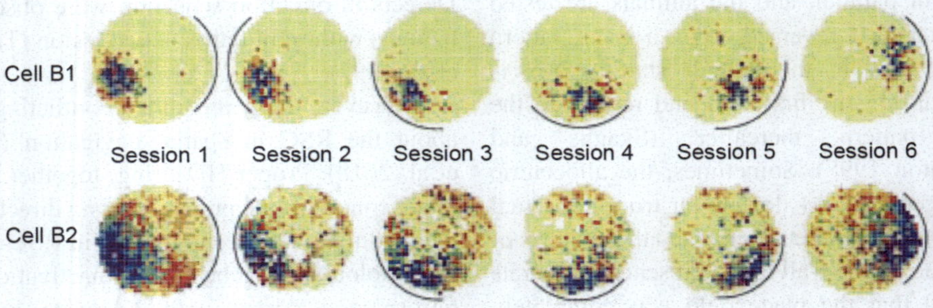

Fig. 7.5 Perceptual representation of the place firing. The two rows of place fields are results from two cells undergoing a series of visible cue card rotations. After combining visible 180° and 45° cue card rotation as illustrated from session 1 to session 6, the place field remaps into a new location within the identical environment (Alexander and Muller 1997)

et al. 1990; Zhang et al. 2014). Previous experiments have further demonstrated that path integration can only support the maintenance of spatially specific activities, and path integration itself is not enough to produce this spatially specific firing pattern. For example, the directional firing of HD cells is abolished if rats are initially placed in a maze in darkness as HD cells need a certain period of light illumination to restore directional firing (Mizumori and Williams 1993). Therefore, if the spatially specific firing patterns reflect the allocentric space perception as discussed above, what is the relationship between path integration and allocentric space perception? Under which conditions does path integration have a similar or identical effect on spatially specific firing patterns?

We argue that path integration is essentially the allocentric space perception driven by idiothetic information. The performance difference between path integration with the conventional allocentric space perception with visual inputs is due to the distinct properties of the information input. Take the discharge of place cells with the light on and off as an example. When an animal moves within an environment, it receives both visual and idiothetic inputs with the lights on. However, in the dark, the animal will have to rely solely on the idiothetic input. Compared to the visual input, which could provide information reflecting distance and angles relative to the environment, idiothetic input has limitations in nature. The idiothetic, typically describing vestibular input about motion, is triggered only during acceleration or deceleration and therefore does not encode movement with constant velocity. More importantly, all idiothetic inputs contain no information about the surrounding environment. Therefore, path integration, or allocentric space perception, provided with only idiothetic input, cannot generate similar content of allocentric space perception provided with visual inputs until appended to it (Mcnaughton et al. 1989; Mizumori and Williams 1993; Quirk et al. 1990). Consistent with the place-specific firing pattern in blind rats, path integration, or allocentric space perception provided only with idiothetic input, can also append to the content of allocentric space perception provided with tactile information (Rochefort et al. 2011). Thus, as two of the basic navigation strategies, path integration and map-based navigation have shared mechanisms working on sensory inputs from different modalities.

The assumption above may be the most rational though there could be other theories to explain the relationship between path integration and allocentric perception. Moreover, this hypothesis can also explain the shifts in place fields or other types of spatially specific firing patterns when the animals stay in the dark for long and the necessity to correct errors in the light period (Gallistel 1993). As path integration is present in a broad range of animals, there is no data support to extend the current argument to path integration in other species, such as insects (Heinze et al. 2018; Pfeiffer and Homberg 2014). While birds are able to conduct path integration, and there are certain types of spatially specific firing patterns in the hippocampus (Mittelstaedt and Mittelstaedt 1982; Sherry et al. 2017), more experiments are needed to discuss this topic in birds.

What needs to be emphasized is that the normal vestibular information, which is an important component of the idiothetic input, seems to be required for the firing of both place cells and HD cells along with the proper visual inputs (Russell et al. 2003; Stackman et al. 2002). The significance of these results is unclear as rats with vestibular lesions are able to return to home base quickly under lighting conditions (Wallace et al. 2002).

7.4.3 Allocentric Space Perception in Place Field Repetition

In the allocentric space perception theory, each place-specific firing pattern is the manifestation of the allocentric space perception about the environment space. Thus, the place field repetition observed in the parallel arranged compartments indicates that there are a certain number of independent environments with identical spatial configurations perceived by the animals (Spiers et al. 2015). On the other hand, the disappearance of place field repetition in the radially packed compartments shows that there may be only one or two independent environments perceived by the animals (Grieves et al. 2016).

How could different configurations affect space perception? The different overlap between the corridors connected to the neighboring compartments under the parallel or radially packed compartments may be the cause. The overlap or the shared space may allow the animals to watch into the neighboring compartments and treat them as integral parts of a large environment. Instead, when there is no shared space to allow the animals to simultaneously perceive the neighboring compartments, such as the situation with parallel packed compartments, each compartment may likely be recognized by the animals as an independent environment, with the firing of place cells and grid cells being driven by local cues (Derdikman et al. 2009; Krupic et al. 2015). In this way, the repetition firing pattern observed in parallel packed compartments is an incidental phenomenon that occurs only with identical compartments but not compartments with different sizes (Derdikman et al. 2009; Spiers et al. 2015). Similarly, when the enclosure of two identical boxes is reoriented from the initial parallel design and directly abutted, the repetition of the place field disappears and remaps into a single place field in most trials (Fuhs et al. 2005). In another design, the same phenomenon as the grid firing was obtained when the wall between two compartments was removed (Wernle et al. 2018). Consistent with the discussion above, the initially replicated grid firing pattern in two identical compartments packed side by side could be turned into a single, continuous representation spanning the two compartments after prolonged experience (Carpenter et al. 2015), indicating that the rats are able to perceive the two independent compartments as integral parts of a large environment after extensive training. While the exact role of HD firing remains unclear, HD firing seems to facilitate the proper recognition of local compartments as part of the large environment. The probability of having more place-field repetition in the radial packed multicompartment is significantly increased with LMN lesions (Harland et al. 2017).

7.4.4 Spatial and Nonspatial Information Processing in the Hippocampal Pathway

The allocentric space perception about the environment is not the only cognitive signal processed in the hippocampal pathways. A new set of CA1 cells has been found to discharge at fixed locations and directions relative to objects inside the environment (Deshmukh and Knierim 2013). There are object-vector cells in the mEC that were found to encode similar vector information relative to spatially confined objects (Høydal et al.

2019). When the animals have to consider the other's location, such 'thinking' about the location of the other's position is also identified in the hippocampus (Danjo et al. 2018; Duvelle and Jeffery 2018). In addition to thinking of the other animals, a population of hippocampal CA1 cells was found to encode the distance and direction information relative to the goal location, where the animals rest and get fed (Sarel et al. 2017). More intriguing about these goal-specific firing patterns is that they are memory-based. Similar to goal cells, a small population of cells in CA1 and the subiculum was found to discharge near the rewards (Gauthier and Tank 2018). In the CA1 of the ventral hippocampus, neurons with similar goal-directed firing were found to specifically project to the nucleus accumbens (Ciocchi et al. 2015). The same neurons are capable of encoding both the spatial and nonspatial cognitive signal as discovered in both the hippocampus and EC in a complex task (Aronov et al. 2017). In addition to the location representation, more abstract time representation has also been discovered in the time cells in hippocampus CA1 (MacDonald et al. 2011; Salz et al. 2016). In humans, a remarkable subset of medial temporal lobe (MTL) neurons are selectively activated when the subjects watch different pictures of given individuals, landmarks, or objects (Quiroga et al. 2005). Taken together, spatial and nonspatial cognitive signals are all observed in the hippocampal pathway, and thus the hippocampus has been suggested to be a general map of cognition (Bellmund et al. 2018; Lisman et al. 2017; Schiller et al. 2015). However, the potential external source of information for place cells, grid cells, etc. promotes another possibility that all this cognitive spatial and nonspatial information processing observed in the hippocampal pathway may also be externally imported. This possibility does not exclude complicated information processing and interactions within the hippocampal pathway.

It remains unclear whether the spatial and nonspatial information is processed using the same or differential computational algorithms in the hippocampal pathway. There is an opinion from the computational point of view that the hippocampus may be "blind" to the information inputs by processing the input messages the same way irrespective of their modality and nature (Lisman et al. 2017). However, this suggestion is not aligned with the experiments. First, the mEC and lEC have different prioritizations among the allocentric and egocentric spatial information together with the nonspatial information (Hargreaves et al. 2005; Keene et al. 2016; Knierim et al. 2014; Lisman 2007; Wang et al. 2018). For example, the mEC is specifically required for a visual scene-based spatial task, while the lEC is particularly involved when the animals need to push or dig into a jar for a food reward in the context of the same visual scene (Yoo and Lee 2017). Wang et al. found lEC neurons showing typical egocentric responses to the border (Wang et al. 2018). Second, within the hippocampus trisynaptic pathway, the path from MECII to CA1 via DG and CA3 is thought to be involved typically in spatial information processing (van Strien et al. 2009; Zhang et al. 2013), the path from MECIII to CA1 is crucial for temporal association memory (Kitamura et al. 2015a; Remondes and Schuman 2004; Suh et al. 2011). Recording directly from the neuronal afferents projecting from MECII to DG or from MECIII to CA1 respectively with the fiber photometry demonstrated that visual-cue-dependent persistent activity develops only in the MECII pathway of the freely behaving mice during the learning (Qin et al. 2018) (Fig. 7.6a). The inhibition of the MECII-DG activity disrupts the navigation task in the Morris water maze, whereas inhibition of the MECIII-CA1 projection is ineffective. Meanwhile, selective inhibition of the synaptic transmission at the MECIII to CA1 synapses, but not the path from MECII to CA1 via DG and CA3 with the genomic methods, leads to significant impairments of the temporal association memory (Suh et al. 2011) (Fig. 7.6b). More interesting is that the island cells, which are a small population of neurons surrounding the ocean cells in MECII and contain similar proportions of grid cells as the ocean cells (Sun et al. 2015), project directly to GABAergic interneurons in stratum lacunosum of CA1 and is indifferent to context-specific encoding but indispensable for the temporal association memory (Kitamura et al. 2014, 2015b; Ray et al. 2014).

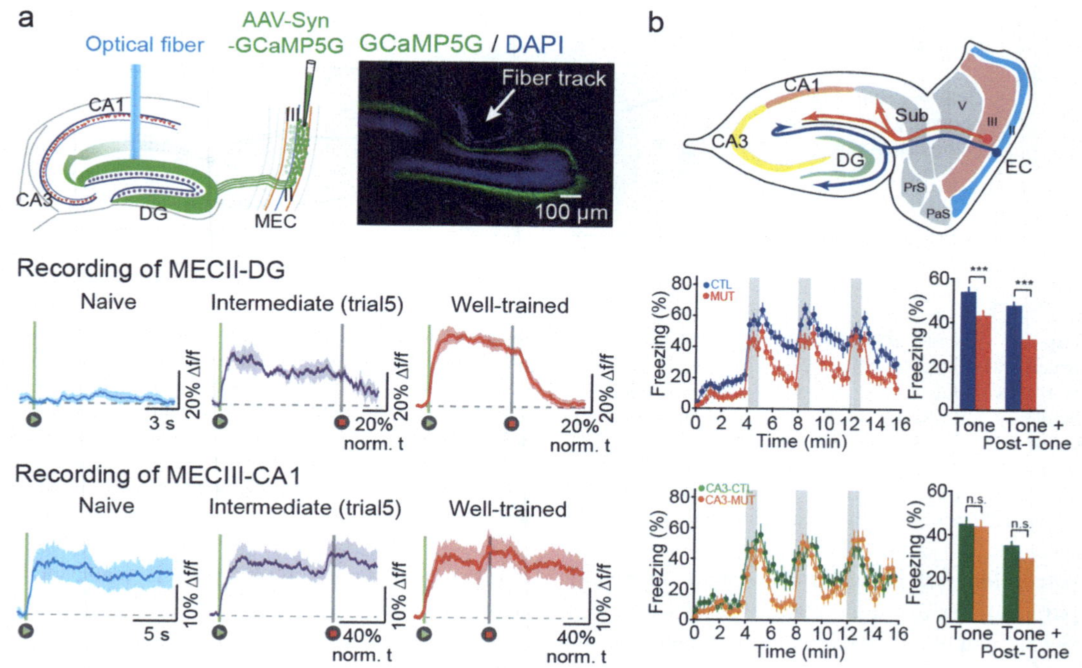

Fig. 7.6 Differential involvement of hippocampus trisynaptic pathway in spatial and nonspatial task. (**a**) Persistent-task-associated activity induced solely in the MECII-DG pathway. Top row: illustration of the fiber photometry recording and histology of fiber recording in axons from MECII to DG. Middle row: Ca^{2+} signal traces during the naive (left), intermediate (middle), and well-trained (right) states recorded from the MECII to DG pathway. Bottom row is the correspondingly Ca^{2+} signal traces recorded from MCIII-CA1 pathway during the naive, intermediate, and well-trained states (Qin et al. 2018). (**b**) Distinct performance of MCIII-CA1 projection in temporal association memory task. Top figure is the illustration of MCIII-CA1, MCII-CA1 via DG and CA3 pathways. The second and third rows are the freezing level measured in mutant and the corresponding control mice on test day (the second day after training) where the synaptic transmission from MECIII to CA1 and CA3 to CA1 are inhibited, respectively (Suh et al. 2011). In both groups all the animals were raised on a doxycycline (Dox) containing diet for 10–12 weeks followed by 4 weeks of a Dox-free diet before the training and test. On the second row, the red line is the result from the transgenic mice with the synaptic transmission inhibited at the MECIII to CA1 synapse (MUT), and the blue line is from the control animals (CTL). On the third row, the orange and blue lines are the freezing level detected in transgenic mice with the synaptic transmission inhibited at the CA3 to CA1 synapse (CA3-MUT) and the corresponding control (CA3-CTL). The right panels on both the second and third rows summarize the freezing levels during the 60-s tone period and the entire 240-s period over the tone and the three first 60-s post-tone duration, respectively

7.5 Conclusion

Since the initial discovery of place cells in the hippocampus, a population of specialized cells representing distinct spatial information has been discovered in a broad range of circuits beyond the hippocampus (Fig. 7.7). Among these spatially specific activities, the shared and distinct properties of place cells, HD cells, and grid cells suggest the possibility that there may be a common mechanism behind the spatial firing of all these cells, which we speculate to be the allocentric space perception or mental representation of the environment as Tolman initially proposed. This allocentric space perception is generated outside and independent of the hippocampus though the neural circuits underlying the allocentric space perception remain unknown. The hippocampus is further hypothesized to receive information from the allocentric space

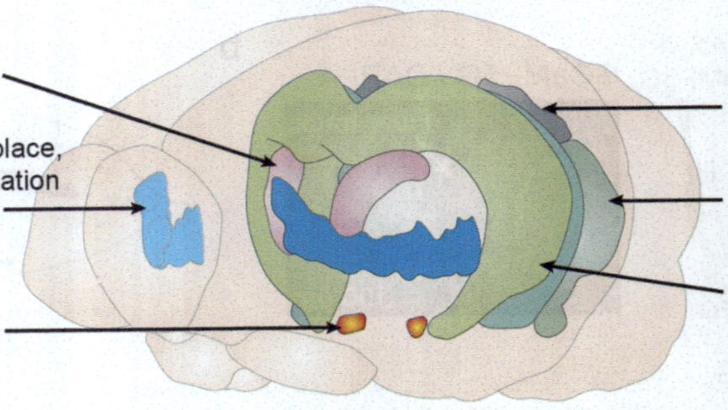

Fig. 7.7 Anatomical locations of allocentric spatial firing patterns within the brain. Here, only place cells, grid cells, HD cells, border cells, and cells encoding place and boundary information in the claustrum are illustrated. Other cells with spatially specific activities, such as the landmark-vector cells in CA1 (Deshmukh and Knierim 2013), object-vector cells in the mEC (Høydal et al. 2019), cells encoding the spatial location of a goal or reward (Gauthier and Tank 2018; Sarel et al. 2017) and together with some HD cells detected in the LDN, RSC, dorsal striatum, and posterior cortex (Chen et al. 1994; Cho and Sharp 2001; Mizumori and Williams 1993; Wiener 1993), are not listed. The location of all the anatomical structures is based on Allen mouse brain atlas and generated in Brain Explorer 2 from the Allen Institute for Brain Science

perception system and process it for mnemonic purposes. As the same allocentric space perception is potentially behind the map-based spatial navigation and path integration, the role of the hippocampus in the spatial navigation using the path integration strategy is also alike, when involved. Thus, when required, the role of the hippocampus in all three navigation strategies is similarly mnemonic.

The mnemonic function of the hippocampus in spatial navigation may present a powerful way to study the mechanisms of memory formation, consolidation, and retrieval (Muller 1996). For example, replays of previous activity in place cells and grid cells observed during sleep and rest have been suggested to be important for memory consolidation (Davidson et al. 2009; Diba and Buzsáki 2007; Foster and Wilson 2006; Gupta et al. 2010; Karlsson and Frank 2009; Lee and Wilson 2002; Louie and Wilson 2001; Ólafsdóttir et al. 2016; Wilson and McNaughton 1994). In another report, the co-occurring patterns during spatial navigation are preserved during the slow-wave sleep (SWS) and rapid-eye-movement (REM) sleep in grid cells (Gardner et al. 2019; Trettel et al. 2019). Meanwhile, the hippocampal circuits may also provide an accessible and efficient platform to study spatial and other nonspatial cognitive functions. For example, sequential firing related to the future path is consistently detected in place cells (Pfeiffer and Foster 2013).

Acknowledgments The authors are grateful to Ms. J. Lou for help in composing the figures. This work was supported by grants from the National Natural Science Foundation of China (No. 31861143038, 81671106) and the Ministry of Science and Technology of China (2015CB759500, 2018YFA0109600). X. Chen is a junior fellow of the CAS Center for Excellence in Brain Science and Intelligence Technology.

References

Ainge JA, Tamosiunaite M, Woergoetter F, Dudchenko PA (2007b) Hippocampal CA1 place cells encode intended destination on a Maze with multiple choice points. J Neurosci 27:9769–9779. https://doi.org/10.1523/JNEUROSCI.2011-07.2007

Ainge JA, van der Meer MAA, Langston RF, Wood ER (2007a) Exploring the role of context-dependent hippocampal activity in spatial alternation behavior. Hippocampus 17:988–1002. https://doi.org/10.1002/hipo.20301

Alexander R, Muller RU (1997) Variable place–cell coupling to a continuously viewed stimulus: evidence that the hippocampus is part of a perceptual system. Philos Trans Royal Soc Lond Ser B: Biol Sci 352:1505–1513. https://doi.org/10.1098/rstb.1997.0137

Alme CB, Miao C, Jezek K, Treves A, Moser EI, Moser M-B (2014) Place cells in the hippocampus: eleven maps for eleven rooms. PNAS 111:18428–18435. https://doi.org/10.1073/pnas.1421056111

Alvarado MC, Bachevalier J (2005) Comparison of the effects of damage to the perirhinal and parahippocampal cortex on transverse patterning and location memory in rhesus macaques. J Neurosci 25:1599–1609. https://doi.org/10.1523/JNEUROSCI.4457-04.2005

Alyan S, McNaughton BL (1999) Hippocampectomized rats are capable of homing by path integration. Behav Neurosci 113:19–31

Aronov D, Nevers R, Tank DW (2017) Mapping of a non-spatial dimension by the hippocampal–entorhinal circuit. Nature 543:719–722. https://doi.org/10.1038/nature21692

Barry C, Hayman R, Burgess N, Jeffery KJ (2007) Experience-dependent rescaling of entorhinal grids. Nat Neurosci 10:682–684. https://doi.org/10.1038/nn1905

Behrens TEJ, Muller TH, Whittington JCR, Mark S, Baram AB, Stachenfeld KL, Kurth-Nelson Z (2018) What is a cognitive map? Organizing knowledge for flexible behavior. Neuron 100:490–509. https://doi.org/10.1016/j.neuron.2018.10.002

Bellistri E, Aguilar J, Brotons-Mas JR, Foffani G, de la Prida LM (2013) Basic properties of somatosensory-evoked responses in the dorsal hippocampus of the rat. J Physiol 591:2667–2686. https://doi.org/10.1113/jphysiol.2013.251892

Bellmund JL, Deuker L, Navarro Schröder T, Doeller CF (2016) Grid-cell representations in mental simulation. eLife 5:e17089. https://doi.org/10.7554/eLife.17089

Bellmund JLS, Gärdenfors P, Moser EI, Doeller CF (2018) Navigating cognition: spatial codes for human thinking. Science 362. https://doi.org/10.1126/science.aat6766

Bickford-Wimer PC, Nagamoto H, Johnson R, Adler LE, Egan M, Rose GM, Freedman R (1990) Auditory sensory gating in hippocampal neurons: a model system in the rat. Biol Psychiatry 27:183–192. https://doi.org/10.1016/0006-3223(90)90648-L

Bjerknes TL, Dagslott NC, Moser EI, Moser M-B (2018) Path integration in place cells of developing rats. Proc Natl Acad Sci U S A 115:E1637–E1646. https://doi.org/10.1073/pnas.1719054115

Bjerknes TL, Langston RF, Kruge IU, Moser EI, Moser M-B (2015) Coherence among head direction cells before eye opening in rat pups. Curr Biol 25:103–108. https://doi.org/10.1016/j.cub.2014.11.009

Bjerknes TL, Moser EI, Moser M-B (2014) Representation of geometric borders in the developing rat. Neuron 82:71–78. https://doi.org/10.1016/j.neuron.2014.02.014

Blair HT, Cho J, Sharp PE (1998) Role of the lateral mammillary nucleus in the rat head direction circuit: a combined single unit recording and lesion study. Neuron 21:1387–1397. https://doi.org/10.1016/S0896-6273(00)80657-1

Blair HT, Sharp PE (1996) Visual and vestibular influences on head-direction cells in the anterior thalamus of the rat. Behav Neurosci 110:643–660. https://doi.org/10.1037/0735-7044.110.4.643

Boccara CN, Sargolini F, Thoresen VH, Solstad T, Witter MP, Moser EI, Moser M-B (2010) Grid cells in pre- and parasubiculum. Nat Neurosci 13:987–994. https://doi.org/10.1038/nn.2602

Bonnevie T, Dunn B, Fyhn M, Hafting T, Derdikman D, Kubie JL, Roudi Y, Moser EI, Moser M-B (2013) Grid cells require excitatory drive from the hippocampus. Nat Neurosci 16:309–317. https://doi.org/10.1038/nn.3311

Bostock E, Muller RU, Kubie JL (1991) Experience-dependent modifications of hippocampal place cell firing. Hippocampus 1:193–205. https://doi.org/10.1002/hipo.450010207

Brandon MP, Koenig J, Leutgeb JK, Leutgeb S (2014) New and distinct hippocampal place codes are generated in a new environment during septal inactivation. Neuron 82:789–796. https://doi.org/10.1016/j.neuron.2014.04.013

Brown JE, Yates BJ, Taube JS (2002) Does the vestibular system contribute to head direction cell activity in the rat? Physiol Behav 77:743–748. https://doi.org/10.1016/S0031-9384(02)00928-9

Brun VH, Leutgeb S, Wu H-Q, Schwarcz R, Witter MP, Moser EI, Moser M-B (2008a) Impaired spatial representation in CA1 after lesion of direct input from entorhinal cortex. Neuron 57:290–302. https://doi.org/10.1016/j.neuron.2007.11.034

Brun VH, Otnæss MK, Molden S, Steffenach H-A, Witter MP, Moser M-B, Moser EI (2002) Place cells and place recognition maintained by direct entorhinal-hippocampal circuitry. Science 296:2243–2246. https://doi.org/10.1126/science.1071089

Brun VH, Solstad T, Kjelstrup KB, Fyhn M, Witter MP, Moser EI, Moser M-B (2008b) Progressive increase in grid scale from dorsal to ventral medial entorhinal cortex. Hippocampus 18:1200–1212. https://doi.org/10.1002/hipo.20504

Buzsáki G, Moser EI (2013) Memory, navigation and theta rhythm in the hippocampal-entorhinal system. Nat Neurosci 16:130–138. https://doi.org/10.1038/nn.3304

Calton JL, Stackman RW, Goodridge JP, Archey WB, Dudchenko PA, Taube JS (2003) Hippocampal place cell instability after lesions of the head direction cell network. J Neurosci 23:9719–9731. https://doi.org/10.1523/JNEUROSCI.23-30-09719.2003

Calton JL, Turner CS, Cyrenne D-LM, Lee BR, Taube JS (2008) Landmark control and updating of self-movement cues are largely maintained in head direction cells after lesions of the posterior parietal cortex. Behav Neurosci 122:827–840. https://doi.org/10.1037/0735-7044.122.4.827

Carpenter F, Manson D, Jeffery K, Burgess N, Barry C (2015) Grid cells form a global representation of connected environments. Curr Biol 25:1176–1182. https://doi.org/10.1016/j.cub.2015.02.037

Cartwright BA, Collett TS (1982) How honey bees use landmarks to guide their return to a food source. Nature 295:560–564. https://doi.org/10.1038/295560a0

Chen G, King JA, Lu Y, Cacucci F, Burgess N (2018) Spatial cell firing during virtual navigation of open arenas by head-restrained mice. eLife 7:e34789. https://doi.org/10.7554/eLife.34789

Chen G, Lu Y, King JA, Cacucci F, Burgess N (2019) Differential influences of environment and self-motion on place and grid cell firing. Nat Commun 10:630. https://doi.org/10.1038/s41467-019-08550-1

Chen G, Manson D, Cacucci F, Wills TJ (2016) Absence of visual input results in the disruption of grid cell firing in the mouse. Curr Biol 26:2335–2342. https://doi.org/10.1016/j.cub.2016.06.043

Chen LL, Lin LH, Green EJ, Barnes CA, McNaughton BL (1994) Head-direction cells in the rat posterior cortex. I. Anatomical distribution and behavioral modulation. Exp Brain Res 101:8–23

Cho J, Sharp PE (2001) Head direction, place, and movement correlates for cells in the rat retrosplenial cortex. Behav Neurosci 115:3–25

Ciocchi S, Passecker J, Malagon-Vina H, Mikus N, Klausberger T (2015) Selective information routing by ventral hippocampal CA1 projection neurons. Science 348:560–563. https://doi.org/10.1126/science.aaa3245

Clark BJ, Simmons CM, Berkowitz LE, Wilber AA (2018) The retrosplenial-parietal network and reference frame coordination for spatial navigation. Behav Neurosci 132:416–429. https://doi.org/10.1037/bne0000260

Clark BJ, Taube JS (2011) Intact landmark control and angular path integration by head direction cells in the anterodorsal thalamus after lesions of the medial entorhinal cortex. Hippocampus 21:767–782. https://doi.org/10.1002/hipo.20874

Cohen NJ, Eichenbaum H (1993) Memory, amnesia, and the hippocampal system, memory, amnesia, and the hippocampal system. The MIT Press, Cambridge, MA

Colgin LL, Moser EI, Moser M-B (2008) Understanding memory through hippocampal remapping. Trends Neurosci 31:469–477. https://doi.org/10.1016/j.tins.2008.06.008

Collett TS, Collett M (2002) Memory use in insect visual navigation. Nat Rev Neurosci 3:542–552. https://doi.org/10.1038/nrn872

Constantinescu AO, O'Reilly JX, Behrens TEJ (2016) Organizing conceptual knowledge in humans with a gridlike code. Science 352:1464–1468. https://doi.org/10.1126/science.aaf0941

Cooper BG, Mizumori SJY (1999) Retrosplenial cortex inactivation selectively impairs navigation in darkness. Neuroreport 10:625–630

Cooper BG, Mizumori SJY (2001) Temporary inactivation of the retrosplenial cortex causes a transient reorganization of spatial coding in the hippocampus. J Neurosci 21:3986–4001. https://doi.org/10.1523/JNEUROSCI.21-11-03986.2001

Danjo T, Toyoizumi T, Fujisawa S (2018) Spatial representations of self and other in the hippocampus. Science 359:213–218. https://doi.org/10.1126/science.aao3898

Davidson TJ, Kloosterman F, Wilson MA (2009) Hippocampal replay of extended experience. Neuron 63:497–507. https://doi.org/10.1016/j.neuron.2009.07.027

De Nigris A, Piccardi L, Bianchini F, Palermo L, Incoccia C, Guariglia C (2013) Role of visuo-spatial working memory in path integration disorders in neglect. Cortex 49:920–930. https://doi.org/10.1016/j.cortex.2012.03.009

Derdikman D, Whitlock JR, Tsao A, Fyhn M, Hafting T, Moser M-B, Moser EI (2009) Fragmentation of grid cell maps in a multicompartment environment. Nat Neurosci 12:1325–1332. https://doi.org/10.1038/nn.2396

Deshmukh SS, Knierim JJ (2013) Influence of local objects on hippocampal representations: landmark vectors and memory. Hippocampus 23:253–267. https://doi.org/10.1002/hipo.22101

Devan BD, McDonald RJ, White NM (1999) Effects of medial and lateral caudate-putamen lesions on place- and cue-guided behaviors in the water maze: relation to thigmotaxis. Behav Brain Res 100:5–14. https://doi.org/10.1016/S0166-4328(98)00107-7

Diba K, Buzsáki G (2007) Forward and reverse hippocampal place-cell sequences during ripples. Nat Neurosci 10:1241–1242. https://doi.org/10.1038/nn1961

Doeller CF, Barry C, Burgess N (2010) Evidence for grid cells in a human memory network. Nature 463:657. https://doi.org/10.1038/nature08704

Dombeck DA, Harvey CD, Tian L, Looger LL, Tank DW (2010) Functional imaging of hippocampal place cells at cellular resolution during virtual navigation. Nat Neurosci 13:1433–1440. https://doi.org/10.1038/nn.2648

Dumont JR, Taube JS (2015) The neural correlates of navigation beyond the hippocampus. Prog Brain Res 219:83–102. https://doi.org/10.1016/bs.pbr.2015.03.004

Duvelle É, Jeffery KJ (2018) Social spaces: place cells represent the locations of others. Curr Biol 28:R271–R273. https://doi.org/10.1016/j.cub.2018.02.017

Eichenbaum H (2000) Hippocampus: mapping or memory? Curr Biol 10:R785–R787. https://doi.org/10.1016/S0960-9822(00)00763-6

Eichenbaum H (2017) The role of the hippocampus in navigation is memory. J Neurophysiol 117:1785–1796. https://doi.org/10.1152/jn.00005.2017

Eichenbaum H, Amaral DG, Buffalo EA, Buzsáki G, Cohen N, Davachi L, Frank L, Heckers S, Morris RGM, Moser EI, Nadel L, O'Keefe J, Preston A, Ranganath C, Silva A, Witter M (2016) Hippocampus at 25. Hippocampus 26:1238–1249. https://doi.org/10.1002/hipo.22616

Eichenbaum H, Cohen NJ (2014) Can we reconcile the declarative memory and spatial navigation views on hippocampal function? Neuron 83:764–770. https://doi.org/10.1016/j.neuron.2014.07.032

Eichenbaum H, Yonelinas AP, Ranganath C (2007) The medial temporal lobe and recognition memory. Annu Rev Neurosci 30:123–152. https://doi.org/10.1146/annurev.neuro.30.051606.094328

Ekstrom AD, Kahana MJ, Caplan JB, Fields TA, Isham EA, Newman EL, Fried I (2003) Cellular networks underlying human spatial navigation. Nature 425:184–188. https://doi.org/10.1038/nature01964

Epstein RA (2008) Parahippocampal and retrosplenial contributions to human spatial navigation. Trends Cogn Sci (Regul Ed) 12:388–396. https://doi.org/10.1016/j.tics.2008.07.004

Etienne AS, Jeffery KJ (2004) Path integration in mammals. Hippocampus 14:180–192. https://doi.org/10.1002/hipo.10173

Foster DJ, Wilson MA (2006) Reverse replay of behavioural sequences in hippocampal place cells during the awake state. Nature 440:680. https://doi.org/10.1038/nature04587

Foster TC, Castro CA, McNaughton BL (1989) Spatial selectivity of rat hippocampal neurons: dependence on preparedness for movement. Science 244:1580–1582. https://doi.org/10.1126/science.2740902

Fuhs MC, Van Rhoads SR, Casale AE, McNaughton B, Touretzky DS (2005) Influence of path integration versus environmental orientation on place cell remapping between visually identical environments. J Neurophysiol 94:2603–2616. https://doi.org/10.1152/jn.00132.2005

Fyhn M, Hafting T, Treves A, Moser M-B, Moser EI (2007) Hippocampal remapping and grid realignment in entorhinal cortex. Nature 446:190–194. https://doi.org/10.1038/nature05601

Fyhn M, Hafting T, Witter MP, Moser EI, Moser M-B (2008) Grid cells in mice. Hippocampus 18:1230–1238. https://doi.org/10.1002/hipo.20472

Fyhn M, Molden S, Witter MP, Moser EI, Moser M-B (2004) Spatial representation in the entorhinal cortex. Science 305:1258–1264. https://doi.org/10.1126/science.1099901

Gallistel CR (1993) The organization of learning (learning, development, and conceptual change) by Gallistel Charles R. (1993-03-02) Paperback. A Bradford Book

Gardner RJ, Lu L, Wernle T, Moser M-B, Moser EI (2019) Correlation structure of grid cells is preserved during sleep. Nat Neurosci 22:598. https://doi.org/10.1038/s41593-019-0360-0

Gauthier JL, Tank DW (2018) A dedicated population for reward coding in the hippocampus. Neuron 99:179–193 .e7. https://doi.org/10.1016/j.neuron.2018.06.008

Goldfarb EV, Chun MM, Phelps EA (2016) Memory-guided attention: independent contributions of the hippocampus and striatum. Neuron 89:317–324. https://doi.org/10.1016/j.neuron.2015.12.014

Golob EJ, Taube JS (1997) Head direction cells and episodic spatial information in rats without a hippocampus. Proc Natl Acad Sci U S A 94:7645–7650

Golob EJ, Taube JS (1999) Head direction cells in rats with hippocampal or overlying neocortical lesions: evidence for impaired angular path integration. J Neurosci 19:7198–7211. https://doi.org/10.1523/JNEUROSCI.19-16-07198.1999

Goodridge JP, Dudchenko PA, Worboys KA, Golob EJ, Taube JS (1998) Cue control and head direction cells. Behav Neurosci 112:749–761. https://doi.org/10.1037/0735-7044.112.4.749

Goodridge JP, Taube JS (1995) Preferential use of the landmark navigational system by head direction cells in rats. Behav Neurosci 109:49–61. https://doi.org/10.1037/0735-7044.109.1.49

Goodridge JP, Taube JS (1997) Interaction between the postsubiculum and anterior thalamus in the generation of head direction cell activity. J Neurosci 17:9315–9330. https://doi.org/10.1523/JNEUROSCI.17-23-09315.1997

Gothard KM, Skaggs WE, Moore KM, McNaughton BL (1996) Binding of hippocampal CA1 neural activity to multiple reference frames in a landmark-based navigation task. J Neurosci 16:823–835. https://doi.org/10.1523/JNEUROSCI.16-02-00823.1996

Grasby K, Talk A (2013) The anterior claustrum and spatial reversal learning in rats. Brain Res 1499:43–52. https://doi.org/10.1016/j.brainres.2013.01.014

Grieves RM, Jeffery KJ (2017) The representation of space in the brain. Behav Process 135:113–131. https://doi.org/10.1016/j.beproc.2016.12.012

Grieves RM, Jenkins BW, Harland BC, Wood ER, Dudchenko PA (2016) Place field repetition and spatial learning in a multicompartment environment. Hippocampus 26:118–134. https://doi.org/10.1002/hipo.22496

Gupta AS, van der Meer MAA, Touretzky DS, Redish AD (2010) Hippocampal replay is not a simple function of experience. Neuron 65:695–705. https://doi.org/10.1016/j.neuron.2010.01.034

Hafting T, Fyhn M, Molden S, Moser M-B, Moser EI (2005) Microstructure of a spatial map in the entorhinal cortex. Nature 436:801–806. https://doi.org/10.1038/nature03721

Hales JB, Schlesiger MI, Leutgeb JK, Squire LR, Leutgeb S, Clark RE (2014) Medial entorhinal cortex lesions only partially disrupt hippocampal place cells and hippocampus-dependent place memory. Cell Rep 9:893–901. https://doi.org/10.1016/j.celrep.2014.10.009

Hargreaves EL, Rao G, Lee I, Knierim JJ (2005) Major dissociation between medial and lateral entorhinal input to dorsal hippocampus. Science 308:1792–1794. https://doi.org/10.1126/science.1110449

Harland B, Grieves RM, Bett D, Stentiford R, Wood ER, Dudchenko PA (2017) Lesions of the head direction cell system increase hippocampal place field repetition. Curr Biol 27:2706–2712 .e2. https://doi.org/10.1016/j.cub.2017.07.071

Hartley T, Burgess N, Lever C, Cacucci F, O'Keefe J (2000) Modeling place fields in terms of the cortical inputs to the hippocampus. Hippocampus 10:369–379. https://doi.org/10.1002/1098-1063(2000)10:4<369::AID-HIPO3>3.0.CO;2-0

Hartley T, Maguire EA, Spiers HJ, Burgess N (2003) The well-worn route and the path less traveled: distinct neural bases of route following and wayfinding in humans. Neuron 37:877–888. https://doi.org/10.1016/S0896-6273(03)00095-3

Heinze S, Narendra A, Cheung A (2018) Principles of insect path integration. Curr Biol 28:R1043–R1058. https://doi.org/10.1016/j.cub.2018.04.058

Henriksen EJ, Colgin LL, Barnes CA, Witter MP, Moser M-B, Moser EI (2010) Spatial representation along the proximodistal axis of CA1. Neuron 68:127–137. https://doi.org/10.1016/j.neuron.2010.08.042

Heys JG, Rangarajan KV, Dombeck DA (2014) The functional micro-organization of grid cells revealed by cellular-resolution imaging. Neuron 84:1079–1090. https://doi.org/10.1016/j.neuron.2014.10.048

Hill AJ, Best PJ (1981) Effects of deafness and blindness on the spatial correlates of hippocampal unit activity in the rat. Exp Neurol 74:204–217. https://doi.org/10.1016/0014-4886(81)90159-X

Horner AJ, Bisby JA, Zotow E, Bush D, Burgess N (2016) Grid-like processing of imagined navigation. Curr Biol 26:842–847. https://doi.org/10.1016/j.cub.2016.01.042

Høydal ØA, Skytøen ER, Andersson SO, Moser M-B, Moser EI (2019) Object-vector coding in the medial entorhinal cortex. Nature 568:400. https://doi.org/10.1038/s41586-019-1077-7

Jacob P-Y, Poucet B, Liberge M, Save E, Sargolini F (2014) Vestibular control of entorhinal cortex activity in spatial navigation. Front Integr Neurosci 8:38. https://doi.org/10.3389/fnint.2014.00038

Jacobs J, Weidemann CT, Miller JF, Solway A, Burke JF, Wei X-X, Suthana N, Sperling MR, Sharan AD, Fried I, Kahana MJ (2013) Direct recordings of grid-like neuronal activity in human spatial navigation. Nat Neurosci 16:1188–1190. https://doi.org/10.1038/nn.3466

Jankowski MM, O'Mara SM (2015) Dynamics of place, boundary and object encoding in rat anterior claustrum. Front Behav Neurosci 9. https://doi.org/10.3389/fnbeh.2015.00250

Jeffery KJ, Donnett JG, Burgess N, O'Keefe JM (1997) Directional control of hippocampal place fields. Exp Brain Res 117:131–142. https://doi.org/10.1007/s002210050206

John RS, Lawrence E, Vilayanur SR (eds) (2014) The claustrum: structural, functional, and clinical neuroscience, 1st edn. Academic Press, San Diego, CA

Jones MW, Wilson MA (2005) Theta rhythms coordinate hippocampal–prefrontal interactions in a spatial memory task. PLoS Biol 3:e402. https://doi.org/10.1371/journal.pbio.0030402

Julian JB, Keinath AT, Frazzetta G, Epstein RA (2018) Human entorhinal cortex represents visual space using a boundary-anchored grid. Nat Neurosci 21:191. https://doi.org/10.1038/s41593-017-0049-1

Kanter BR, Lykken CM, Avesar D, Weible A, Dickinson J, Dunn B, Borgesius NZ, Roudi Y, Kentros CG (2017) A novel mechanism for the grid-to-place cell transformation revealed by transgenic depolarization of medial entorhinal cortex layer II. Neuron 93:1480–1492 .e6. https://doi.org/10.1016/j.neuron.2017.03.001

Karlsson MP, Frank LM (2009) Awake replay of remote experiences in the hippocampus. Nat Neurosci 12:913–918. https://doi.org/10.1038/nn.2344

Keene CS, Bladon J, McKenzie S, Liu CD, O'Keefe J, Eichenbaum H (2016) Complementary functional organization of neuronal activity patterns in the perirhinal, lateral entorhinal, and medial entorhinal cortices. J Neurosci 36:3660–3675. https://doi.org/10.1523/JNEUROSCI.4368-15.2016

Kentros C, Hargreaves E, Hawkins RD, Kandel ER, Shapiro M, Muller RV (1998) Abolition of long-term stability of new hippocampal place cell maps by NMDA receptor blockade. Science 280:2121–2126. https://doi.org/10.1126/science.280.5372.2121

Kentros CG, Agnihotri NT, Streater S, Hawkins RD, Kandel ER (2004) Increased attention to spatial context increases both place field stability and spatial memory. Neuron 42:283–295. https://doi.org/10.1016/S0896-6273(04)00192-8

Killian NJ, Buffalo EA (2018) Grid cells map the visual world. Nat Neurosci 21:161. https://doi.org/10.1038/s41593-017-0062-4

Killian NJ, Jutras MJ, Buffalo EA (2012) A map of visual space in the primate entorhinal cortex. Nature 491:761–764. https://doi.org/10.1038/nature11587

Kim S, Sapiurka M, Clark RE, Squire LR (2013) Contrasting effects on path integration after hippocampal damage in humans and rats. PNAS 110:4732–4737. https://doi.org/10.1073/pnas.1300869110

Kitamura T, MacDonald CJ, Tonegawa S (2015a) Entorhinal–hippocampal neuronal circuits bridge temporally discontiguous events. Learn Mem 22:438–443. https://doi.org/10.1101/lm.038687.115

Kitamura T, Pignatelli M, Suh J, Kohara K, Yoshiki A, Abe K, Tonegawa S (2014) Island cells control temporal association memory. Science 343:896–901. https://doi.org/10.1126/science.1244634

Kitamura T, Sun C, Martin J, Kitch LJ, Schnitzer MJ, Tonegawa S (2015b) Entorhinal cortical ocean cells encode specific contexts and drive context-specific fear memory. Neuron 87:1317–1331. https://doi.org/10.1016/j.neuron.2015.08.036

Kitanishi T, Matsuo N (2017) Organization of the claustrum-to-entorhinal cortical connection in mice. J

Neurosci 37:269–280. https://doi.org/10.1523/JNEUROSCI.1360-16.2016
Knierim JJ, Kudrimoti HS, McNaughton BL (1995) Place cells, head direction cells, and the learning of landmark stability. J Neurosci 15:1648–1659. https://doi.org/10.1523/JNEUROSCI.15-03-01648.1995
Knierim JJ, Neunuebel JP, Deshmukh SS (2014) Functional correlates of the lateral and medial entorhinal cortex: objects, path integration and local-global reference frames. Philos Trans R Soc Lond B Biol Sci 369:20130369. https://doi.org/10.1098/rstb.2013.0369
Koenig J, Linder AN, Leutgeb JK, Leutgeb S (2011) The spatial periodicity of grid cells is not sustained during reduced theta oscillations. Science 332:592–595. https://doi.org/10.1126/science.1201685
Kononenko NL, Witter MP (2012) Presubiculum layer III conveys retrosplenial input to the medial entorhinal cortex. Hippocampus 22:881–895. https://doi.org/10.1002/hipo.20949
Kropff E, Carmichael JE, Moser M-B, Moser EI (2015) Speed cells in the medial entorhinal cortex. Nature 523:419–424. https://doi.org/10.1038/nature14622
Krupic J, Bauza M, Burton S, Barry C, O'Keefe J (2015) Grid cell symmetry is shaped by environmental geometry. Nature 518:232–235. https://doi.org/10.1038/nature14153
Krupic J, Bauza M, Burton S, O'Keefe J (2016) Framing the grid: effect of boundaries on grid cells and navigation. J Physiol 594:6489–6499. https://doi.org/10.1113/JP270607
Krupic J, Bauza M, Burton S, O'Keefe J (2018) Local transformations of the hippocampal cognitive map. Science 359:1143–1146. https://doi.org/10.1126/science.aao4960
Lambrey S, Doeller C, Berthoz A, Burgess N (2012) Imagining being somewhere else: neural basis of changing perspective in space. Cereb Cortex 22:166–174. https://doi.org/10.1093/cercor/bhr101
Langston RF, Ainge JA, Couey JJ, Canto CB, Bjerknes TL, Witter MP, Moser EI, Moser M-B (2010) Development of the spatial representation system in the rat. Science 328:1576–1580. https://doi.org/10.1126/science.1188210
Lee AK, Wilson MA (2002) Memory of sequential experience in the hippocampus during slow wave sleep. Neuron 36:1183–1194. https://doi.org/10.1016/S0896-6273(02)01096-6
Lee I, Griffin AL, Zilli EA, Eichenbaum H, Hasselmo ME (2006) Gradual translocation of spatial correlates of neuronal firing in the hippocampus toward prospective reward locations. Neuron 51:639–650. https://doi.org/10.1016/j.neuron.2006.06.033
Leszczynski M (2011) How does hippocampus contribute to working memory processing? Front Hum Neurosci 5. https://doi.org/10.3389/fnhum.2011.00168
Leutgeb S, Leutgeb JK, Barnes CA, Moser EI, McNaughton BL, Moser M-B (2005) Independent codes for spatial and episodic memory in hippocampal neuronal ensembles. Science 309:619–623. https://doi.org/10.1126/science.1114037
Leutgeb S, Leutgeb JK, Treves A, Moser M-B, Moser EI (2004) Distinct ensemble codes in hippocampal areas CA3 and CA1. Science 305:1295–1298. https://doi.org/10.1126/science.1100265
Lever C, Burton S, Jeewajee A, O'Keefe J, Burgess N (2009) Boundary vector cells in the subiculum of the hippocampal formation. J Neurosci 29:9771–9777. https://doi.org/10.1523/JNEUROSCI.1319-09.2009
Lisman J, Buzsáki G, Eichenbaum H, Nadel L, Ranganath C, Redish AD (2017) Viewpoints: how the hippocampus contributes to memory, navigation and cognition. Nat Neurosci 20:1434–1447. https://doi.org/10.1038/nn.4661
Lisman JE (2007) Role of the dual entorhinal inputs to hippocampus: a hypothesis based on cue/action (non-self/self) couplets. In: Scharfman HE (ed) Progress in brain research, the dentate gyrus: a comprehensive guide to structure, function, and clinical implications. Elsevier, pp 615–818. https://doi.org/10.1016/S0079-6123(07)63033-7
Lopez C, Blanke O (2011) The thalamocortical vestibular system in animals and humans. Brain Res Rev 67:119–146. https://doi.org/10.1016/j.brainresrev.2010.12.002
Louie K, Wilson MA (2001) Temporally structured replay of awake hippocampal ensemble activity during rapid eye movement sleep. Neuron 29:145–156. https://doi.org/10.1016/S0896-6273(01)00186-6
Maaswinkel H, Jarrard LE, Whishaw IQ (1999) Hippocampectomized rats are impaired in homing by path integration. Hippocampus 9:553–561. https://doi.org/10.1002/(SICI)1098-1063(1999)9:5<553::AID-HIPO9>3.0.CO;2-G
MacDonald CJ, Lepage KQ, Eden UT, Eichenbaum H (2011) Hippocampal "Time Cells" bridge the gap in memory for discontiguous events. Neuron 71:737–749. https://doi.org/10.1016/j.neuron.2011.07.012
Maguire EA, Nannery R, Spiers HJ (2006) Navigation around London by a taxi driver with bilateral hippocampal lesions. Brain 129:2894–2907. https://doi.org/10.1093/brain/awl286
Markus EJ, Barnes CA, McNaughton BL, Gladden VL, Skaggs WE (1994) Spatial information content and reliability of hippocampal CA1 neurons: effects of visual input. Hippocampus 4:410–421. https://doi.org/10.1002/hipo.450040404
Markus EJ, Qin YL, Leonard B, Skaggs WE, McNaughton BL, Barnes CA (1995) Interactions between location and task affect the spatial and directional firing of hippocampal neurons. J Neurosci 15:7079–7094. https://doi.org/10.1523/JNEUROSCI.15-11-07079.1995
Matsumura N, Nishijo H, Tamura R, Eifuku S, Endo S, Ono T (1999) Spatial- and task-dependent neuronal responses during real and virtual translocation in the monkey hippocampal formation. J Neurosci

19:2381–2393. https://doi.org/10.1523/JNEUROSCI.19-06-02381.1999

McDonald RJ, White NM (1993) A triple dissociation of memory systems: Hippocampus, amygdala, and dorsal striatum. Behav Neurosci 107:3–22. https://doi.org/10.1037/0735-7044.107.1.3

McDonald RJ, White NM (1994) Parallel information processing in the water maze: Evidence for independent memory systems involving dorsal striatum and hippocampus. Behav Neural Biol 61:260–270. https://doi.org/10.1016/S0163-1047(05)80009-3

McNaughton BL, Barnes CA, Gerrard JL, Gothard K, Jung MW, Knierim JJ, Kudrimoti H, Qin Y, Skaggs WE, Suster M, Weaver KL (1996) Deciphering the hippocampal polyglot: the hippocampus as a path integration system. J Exp Biol 199:173–185

Mcnaughton BL, Leonard B, Chen L (1989) Cortical-hippocampal interactions and cognitive mapping: A hypothesis based on reintegration of the parietal and inferotemporal pathways for visual processing. Psychobiology 17:230–235. https://doi.org/10.1007/BF03337774

Mehlman ML, Winter SS, Valerio S, Taube JS (2018) Functional and anatomical relationships between the medial precentral cortex, dorsal striatum, and head direction cell circuitry. I Record Stud J Neurophysiol 121:350–370. https://doi.org/10.1152/jn.00143.2018

Meshulam L, Gauthier JL, Brody CD, Tank DW, Bialek W (2017) Collective behavior of place and non-place neurons in the hippocampal network. Neuron 96:1178–1191 .e4. https://doi.org/10.1016/j.neuron.2017.10.027

Miao C, Cao Q, Ito HT, Yamahachi H, Witter MP, Moser M-B, Moser EI (2015) Hippocampal remapping after partial inactivation of the medial entorhinal cortex. Neuron 88:590–603. https://doi.org/10.1016/j.neuron.2015.09.051

Miller VM, Best PJ (1980) Spatial correlates of hippocampal unit activity are altered by lesions of the fornix and entorhinal cortex. Brain Res 194:311–323. https://doi.org/10.1016/0006-8993(80)91214-7

Mitchell AS, Czajkowski R, Zhang N, Jeffery K, Nelson AJD (2018) Retrosplenial cortex and its role in spatial cognition. Brain Neurosci Adv 2. https://doi.org/10.1177/2398212818757098

Mittelstaedt H, Mittelstaedt M-L (1982) Homing by path integration. In: Avian navigation, proceedings in life sciences. Springer, Berlin, Heidelberg, pp 290–297. https://doi.org/10.1007/978-3-642-68616-0_29

Mittelstaedt M-L, Mittelstaedt H (1980) Homing by path integration in a mammal. Naturwissenschaften 67:566–567. https://doi.org/10.1007/BF00450672

Mizumori SJ, Williams JD (1993) Directionally selective mnemonic properties of neurons in the lateral dorsal nucleus of the thalamus of rats. J Neurosci 13:4015–4028. https://doi.org/10.1523/JNEUROSCI.13-09-04015.1993

Mizuseki K, Royer S, Diba K, Buzsáki G (2012) Activity dynamics and behavioral correlates of CA3 and CA1 hippocampal pyramidal neurons. Hippocampus 22:1659–1680. https://doi.org/10.1002/hipo.22002

Morris RGM, Garrud P, Rawlins JNP, O'Keefe J (1982) Place navigation impaired in rats with hippocampal lesions. Nature 297:681–683. https://doi.org/10.1038/297681a0

Moser EI, Moser M-B (2008) A metric for space. Hippocampus 18:1142–1156. https://doi.org/10.1002/hipo.20483

Moser EI, Moser M-B, McNaughton BL (2017) Spatial representation in the hippocampal formation: a history. Nat Neurosci 20:1448–1464. https://doi.org/10.1038/nn.4653

Muller R (1996) A quarter of a century of place cells. Neuron 17:813–822. https://doi.org/10.1016/S0896-6273(00)80214-7

Muller RU, Bostock E, Taube JS, Kubie JL (1994) On the directional firing properties of hippocampal place cells. J Neurosci 14:7235–7251

Muller RU, Kubie JL (1987) The effects of changes in the environment on the spatial firing of hippocampal complex-spike cells. J Neurosci 7:1951–1968. https://doi.org/10.1523/JNEUROSCI.07-07-01951.1987

Muller RU, Kubie JL, Ranck JB (1987) Spatial firing patterns of hippocampal complex-spike cells in a fixed environment. J Neurosci 7:1935–1950. https://doi.org/10.1523/JNEUROSCI.07-07-01935.1987

Nau M, Schröder TN, Bellmund JLS, Doeller CF (2018) Hexadirectional coding of visual space in human entorhinal cortex. Nat Neurosci 21:188. https://doi.org/10.1038/s41593-017-0050-8

O'Keefe J (1976) Place units in the hippocampus of the freely moving rat. Exp Neurol 51:78–109. https://doi.org/10.1016/0014-4886(76)90055-8

O'Keefe J (1979) A review of the hippocampal place cells. Prog Neurobiol 13:419–439. https://doi.org/10.1016/0301-0082(79)90005-4

O'Keefe J (1999) Do hippocampal pyramidal cells signal non-spatial as well as spatial information? Hippocampus 9:352–364. https://doi.org/10.1002/(SICI)1098-1063(1999)9:4<352::AID-HIPO3>3.0.CO;2-1

O'Keefe J, Burgess N, Donnett JG, Jeffery KJ, Maguire EA (1998) Place cells, navigational accuracy, and the human hippocampus. Philos Trans R Soc Lond Ser B Biol Sci 353:1333–1340

O'Keefe J, Conway DH (1978) Hippocampal place units in the freely moving rat: why they fire where they fire. Exp Brain Res 31:573–590

O'Keefe J, Dostrovsky J (1971) The hippocampus as a spatial map. Preliminary evidence from unit activity in the freely-moving rat. Brain Res 34:171–175. https://doi.org/10.1016/0006-8993(71)90358-1

O'Keefe J, Speakman A (1987) Single unit activity in the rat hippocampus during a spatial memory task. Exp Brain Res 68:1–27

O'Keefe J, Nadel L (1978) The hippocampus as a cognitive map. Clarendon Press, Oxford

Ólafsdóttir HF, Carpenter F, Barry C (2016) Coordinated grid and place cell replay during rest. Nat Neurosci 19:792–794. https://doi.org/10.1038/nn.4291

Ormond J, McNaughton BL (2015) Place field expansion after focal MEC inactivations is consistent with loss of Fourier components and path integrator gain reduction. Proc Natl Acad Sci U S A 112:4116–4121. https://doi.org/10.1073/pnas.1421963112

Packard MG, Knowlton BJ (2002) Learning and memory functions of the Basal Ganglia. Annu Rev Neurosci 25:563–593. https://doi.org/10.1146/annurev.neuro.25.112701.142937

Packard MG, McGaugh JL (1996) Inactivation of hippocampus or caudate nucleus with lidocaine differentially affects expression of place and response learning. Neurobiol Learn Mem 65:65–72. https://doi.org/10.1006/nlme.1996.0007

Park S, Tyszka M, Allman JM (2012) The claustrum and insula in Microcebus murinus: a High Resolution Diffusion Imaging Study. Front Neuroanat 6. https://doi.org/10.3389/fnana.2012.00021

Parron C, Save E (2004) Evidence for entorhinal and parietal cortices involvement in path integration in the rat. Exp Brain Res 159:349–359. https://doi.org/10.1007/s00221-004-1960-8

Pereira A, Ribeiro S, Wiest M, Moore LC, Pantoja J, Lin S-C, Nicolelis MAL (2007) Processing of tactile information by the hippocampus. PNAS 104:18286–18291. https://doi.org/10.1073/pnas.0708611104

Pérez-Escobar JA, Kornienko O, Latuske P, Kohler L, Allen K (2016) Visual landmarks sharpen grid cell metric and confer context specificity to neurons of the medial entorhinal cortex. eLife Sci 5:e16937. https://doi.org/10.7554/eLife.16937

Pfeiffer BE, Foster DJ (2013) Hippocampal place-cell sequences depict future paths to remembered goals. Nature 497:74–79. https://doi.org/10.1038/nature12112

Pfeiffer K, Homberg U (2014) Organization and functional roles of the central complex in the insect brain. Annu Rev Entomol 59:165–184. https://doi.org/10.1146/annurev-ento-011613-162031

Poucet B, Benhamou S (1997) The neuropsychology of spatial cognition in the rat. Crit Rev Neurobiol 11:101–120

Poucet B, Save E, Lenck-Santini PP (2000) Sensory and memory properties of hippocampal place cells. Rev Neurosci 11:95–111

Qin H, Fu L, Hu B, Liao X, Lu J, He W, Liang S, Zhang K, Li R, Yao J, Yan J, Chen H, Jia H, Zott B, Konnerth A, Chen X (2018) A visual-cue-dependent memory circuit for place navigation. Neuron 99:47–55 .e4. https://doi.org/10.1016/j.neuron.2018.05.021

Quirk GJ, Muller RU, Kubie JL (1990) The firing of hippocampal place cells in the dark depends on the rat's recent experience. J Neurosci 10:2008–2017. https://doi.org/10.1523/JNEUROSCI.10-06-02008.1990

Quiroga RQ, Reddy L, Kreiman G, Koch C, Fried I (2005) Invariant visual representation by single neurons in the human brain. Nature 435:1102. https://doi.org/10.1038/nature03687

Ranck JB (1984) Head-direction cells in the deep cell layers of dorsal presubiculum in freely moving rats. Soc Neurosci 10:599

Ray S, Naumann R, Burgalossi A, Tang Q, Schmidt H, Brecht M (2014) Grid-layout and theta-modulation of layer 2 pyramidal neurons in medial entorhinal cortex. Science 343:891–896. https://doi.org/10.1126/science.1243028

Redish AD, Rosenzweig ES, Bohanick JD, McNaughton BL, Barnes CA (2000) Dynamics of hippocampal ensemble activity realignment: time versus space. J Neurosci 20:9298–9309. https://doi.org/10.1523/JNEUROSCI.20-24-09298.2000

Remondes M, Schuman EM (2004) Role for a cortical input to hippocampal area CA1 in the consolidation of a long-term memory. Nature 431:699–703. https://doi.org/10.1038/nature02965

Rochefort C, Arabo A, André M, Poucet B, Save E, Rondi-Reig L (2011) Cerebellum shapes hippocampal spatial code. Science 334:385–389. https://doi.org/10.1126/science.1207403

Rolls ET (1999) Spatial view cells and the representation of place in the primate hippocampus. Hippocampus 9:467–480. https://doi.org/10.1002/(SICI)1098-1063(1999)9:4<467::AID-HIPO13>3.0.CO;2-F

Rolls ET, Robertson RG, Georges-François P (1997) Spatial view cells in the primate hippocampus. Eur J Neurosci 9:1789–1794. https://doi.org/10.1111/j.1460-9568.1997.tb01538.x

Rosenbaum RS, Priselac S, Köhler S, Black SE, Gao F, Nadel L, Moscovitch M (2000) Remote spatial memory in an amnesic person with extensive bilateral hippocampal lesions. Nat Neurosci 3:1044–1048. https://doi.org/10.1038/79867

Rueckemann JW, DiMauro AJ, Rangel LM, Han X, Boyden ES, Eichenbaum H (2016) Transient optogenetic inactivation of the medial entorhinal cortex biases the active population of hippocampal neurons. Hippocampus 26:246–260. https://doi.org/10.1002/hipo.22519

Russell NA, Horii A, Smith PF, Darlington CL, Bilkey DK (2003) Long-term effects of permanent vestibular lesions on hippocampal spatial firing. J Neurosci 23:6490–6498. https://doi.org/10.1523/JNEUROSCI.23-16-06490.2003

Salz DM, Tiganj Z, Khasnabish S, Kohley A, Sheehan D, Howard MW, Eichenbaum H (2016) Time cells in hippocampal area CA3. J Neurosci 36:7476–7484. https://doi.org/10.1523/JNEUROSCI.0087-16.2016

Sarel A, Finkelstein A, Las L, Ulanovsky N (2017) Vectorial representation of spatial goals in the hippocampus of bats. Science 355:176–180. https://doi.org/10.1126/science.aak9589

Sargolini F, Fyhn M, Hafting T, McNaughton BL, Witter MP, Moser M-B, Moser EI (2006) Conjunctive representation of position, direction, and velocity in entorhinal cortex. Science 312:758–762. https://doi.org/10.1126/science.1125572

Save E, Cressant A, Thinus-Blanc C, Poucet B (1998) Spatial firing of hippocampal place cells in blind rats. J Neurosci 18:1818–1826. https://doi.org/10.1523/JNEUROSCI.18-05-01818.1998

Save E, Nerad L, Poucet B (2000) Contribution of multiple sensory information to place field stability in hippocampal place cells. Hippocampus 10:64–76. https://doi.org/10.1002/(SICI)1098-1063(2000)10:1<64::AID-HIPO7>3.0.CO;2-Y

Savelli F, Yoganarasimha D, Knierim JJ (2008) Influence of boundary removal on the spatial representations of the medial entorhinal cortex. Hippocampus 18:1270–1282. https://doi.org/10.1002/hipo.20511

Schiller D, Eichenbaum H, Buffalo EA, Davachi L, Foster DJ, Leutgeb S, Ranganath C (2015) Memory and space: towards an understanding of the cognitive map. J Neurosci 35:13904–13911. https://doi.org/10.1523/JNEUROSCI.2618-15.2015

Schlesiger MI, Boublil BL, Hales JB, Leutgeb JK, Leutgeb S (2018) Hippocampal global remapping can occur without input from the medial entorhinal cortex. Cell Rep 22:3152–3159. https://doi.org/10.1016/j.celrep.2018.02.082

Schlesiger MI, Cannova CC, Boublil BL, Hales JB, Mankin EA, Brandon MP, Leutgeb JK, Leibold C, Leutgeb S (2015) The medial entorhinal cortex is necessary for temporal organization of hippocampal neuronal activity. Nat Neurosci 18:1123–1132. https://doi.org/10.1038/nn.4056

Seger CA, Spiering BJ (2011) A critical review of habit learning and the Basal Ganglia. Front Syst Neurosci 5. https://doi.org/10.3389/fnsys.2011.00066

Sharp PE, Koester K (2008) Lesions of the mammillary body region severely disrupt the cortical head direction, but not place cell signal. Hippocampus 18:766–784. https://doi.org/10.1002/hipo.20436

Sherry DF, Grella SL, Guigueno MF, White DJ, Marrone DF (2017) Are there place cells in the avian hippocampus? BBE 90:73–80. https://doi.org/10.1159/000477085

Shrager Y, Kirwan CB, Squire LR (2008) Neural basis of the cognitive map: path integration does not require hippocampus or entorhinal cortex. PNAS 105:12034–12038. https://doi.org/10.1073/pnas.0805414105

Skaggs WE, McNaughton BL (1998) Spatial firing properties of hippocampal CA1 populations in an environment containing two visually identical regions. J Neurosci 18:8455–8466. https://doi.org/10.1523/JNEUROSCI.18-20-08455.1998

Snyder LH, Grieve KL, Brotchie P, Andersen RA (1998) Separate body- and world-referenced representations of visual space in parietal cortex. Nature 394:887–891. https://doi.org/10.1038/29777

Solstad T, Boccara CN, Kropff E, Moser M-B, Moser EI (2008) Representation of geometric borders in the entorhinal cortex. Science 322:1865–1868. https://doi.org/10.1126/science.1166466

Spellman T, Rigotti M, Ahmari SE, Fusi S, Gogos JA, Gordon JA (2015) Hippocampal–prefrontal input supports spatial encoding in working memory. Nature 522:309–314. https://doi.org/10.1038/nature14445

Spiers HJ, Hayman RMA, Jovalekic A, Marozzi E, Jeffery KJ (2015) Place field repetition and purely local remapping in a multicompartment environment. Cereb Cortex 25:10–25. https://doi.org/10.1093/cercor/bht198

Squire LR, Zola SM (1996) Structure and function of declarative and nondeclarative memory systems. PNAS 93:13515–13522. https://doi.org/10.1073/pnas.93.24.13515

Stackman RW, Clark AS, Taube JS (2002) Hippocampal spatial representations require vestibular input. Hippocampus 12:291–303. https://doi.org/10.1002/hipo.1112

Stackman RW, Taube JS (1997) Firing properties of head direction cells in the rat anterior thalamic nucleus: dependence on vestibular input. J Neurosci 17:4349–4358. https://doi.org/10.1523/JNEUROSCI.17-11-04349.1997

Stackman RW, Taube JS (1998) Firing properties of rat lateral mammillary single units: head direction, head pitch, and angular head velocity. J Neurosci 18:9020–9037. https://doi.org/10.1523/JNEUROSCI.18-21-09020.1998

Steffenach H-A, Witter M, Moser M-B, Moser EI (2005) Spatial memory in the rat requires the dorsolateral band of the entorhinal cortex. Neuron 45:301–313. https://doi.org/10.1016/j.neuron.2004.12.044

Stein JF (1989) Representation of egocentric space in the posterior parietal cortex. Q J Exp Physiol 74:583–606. https://doi.org/10.1113/expphysiol.1989.sp003314

Stensola H, Stensola T, Solstad T, Frøland K, Moser M-B, Moser EI (2012) The entorhinal grid map is discretized. Nature 492:72–78. https://doi.org/10.1038/nature11649

Stensola T, Stensola H, Moser M-B, Moser EI (2015) Shearing-induced asymmetry in entorhinal grid cells. Nature 518:207–212. https://doi.org/10.1038/nature14151

Suh J, Rivest AJ, Nakashiba T, Tominaga T, Tonegawa S (2011) Entorhinal cortex layer III input to the hippocampus is crucial for temporal association memory. Science 334:1415–1420. https://doi.org/10.1126/science.1210125

Sun C, Kitamura T, Yamamoto J, Martin J, Pignatelli M, Kitch LJ, Schnitzer MJ, Tonegawa S (2015) Distinct speed dependence of entorhinal island and ocean cells, including respective grid cells. PNAS 112:9466–9471. https://doi.org/10.1073/pnas.1511668112

Suthana N, Haneef Z, Stern J, Mukamel R, Behnke E, Knowlton B, Fried I (2012) Memory enhancement and deep-brain stimulation of the entorhinal area. N

Engl J Med 366:502–510. https://doi.org/10.1056/NEJMoa1107212

Takahashi N, Kawamura M, Shiota J, Kasahata N, Hirayama K (1997) Pure topographic disorientation due to right retrosplenial lesion. Neurology 49:464–469

Taube JS (1995) Head direction cells recorded in the anterior thalamic nuclei of freely moving rats. J Neurosci 15:70–86. https://doi.org/10.1523/JNEUROSCI.15-01-00070.1995

Taube JS, Bassett JP (2003) Persistent neural activity in head direction cells. Cereb Cortex 13:1162–1172

Taube JS, Muller RU, Ranck JB (1990a) Head-direction cells recorded from the postsubiculum in freely moving rats. I. Description and quantitative analysis. J Neurosci 10:420–435. https://doi.org/10.1523/JNEUROSCI.10-02-00420.1990

Taube JS, Muller RU, Ranck JB (1990b) Head-direction cells recorded from the postsubiculum in freely moving rats. II Effects of environmental manipulations. J Neurosci 10:436–447. https://doi.org/10.1523/JNEUROSCI.10-02-00436.1990

Teng E, Squire LR (1999) Memory for places learned long ago is intact after hippocampal damage. Nature 400:675–677. https://doi.org/10.1038/23276

Thompson LT, Best PJ (1989) Place cells and silent cells in the hippocampus of freely-behaving rats. J Neurosci 9:2382–2390. https://doi.org/10.1523/JNEUROSCI.09-07-02382.1989

Thompson LT, Best PJ (1990) Long-term stability of the place-field activity of single units recorded from the dorsal hippocampus of freely behaving rats. Brain Res 509:299–308. https://doi.org/10.1016/0006-8993(90)90555-P

Tocker G, Barak O, Derdikman D (2015) Grid cells correlation structure suggests organized feedforward projections into superficial layers of the medial entorhinal cortex. Hippocampus 25:1599–1613. https://doi.org/10.1002/hipo.22481

Tolman EC (1948) Cognitive maps in rats and men. Psychol Rev 55:189–208

Trettel SG, Trimper JB, Hwaun E, Fiete IR, Colgin LL (2019) Grid cell co-activity patterns during sleep reflect spatial overlap of grid fields during active behaviors. Nat Neurosci 22:609. https://doi.org/10.1038/s41593-019-0359-6

Ulanovsky N, Moss CF (2007) Hippocampal cellular and network activity in freely moving echolocating bats. Nat Neurosci 10:224–233. https://doi.org/10.1038/nn1829

Urgolites ZJ, Kim S, Hopkins RO, Squire LR (2016) Map reading, navigating from maps, and the medial temporal lobe. PNAS 113:14289–14293. https://doi.org/10.1073/pnas.1617786113

Van Cauter T, Poucet B, Save E (2008) Unstable CA1 place cell representation in rats with entorhinal cortex lesions. Eur J Neurosci 27:1933–1946. https://doi.org/10.1111/j.1460-9568.2008.06158.x

van Groen T, Wyss JM (1992) Connections of the retrosplenial dysgranular cortex in the rat. J Comp Neurol 315:200–216. https://doi.org/10.1002/cne.903150207

van Strien NM, Cappaert NLM, Witter MP (2009) The anatomy of memory: an interactive overview of the parahippocampal–hippocampal network. Nat Rev Neurosci 10:272–282. https://doi.org/10.1038/nrn2614

Vann SD, Aggleton JP (2005) Selective dysgranular retrosplenial cortex lesions in rats disrupt allocentric performance of the radial-arm maze task. Behav Neurosci 119:1682–1686. https://doi.org/10.1037/0735-7044.119.6.1682

Vann SD, Aggleton JP, Maguire EA (2009) What does the retrosplenial cortex do? Nat Rev Neurosci 10:792–802. https://doi.org/10.1038/nrn2733

Vass LK, Epstein RA (2013) Abstract representations of location and facing direction in the human brain. J Neurosci 33:6133–6142. https://doi.org/10.1523/JNEUROSCI.3873-12.2013

Vogt BA, Miller MW (1983) Cortical connections between rat cingulate cortex and visual, motor, and postsubicular cortices. J Comp Neurol 216:192–210. https://doi.org/10.1002/cne.902160207

Wallace DG, Hines DJ, Pellis SM, Whishaw IQ (2002) Vestibular information is required for dead reckoning in the rat. J Neurosci 22:10009–10017. https://doi.org/10.1523/JNEUROSCI.22-22-10009.2002

Wang C, Chen X, Lee H, Deshmukh SS, Yoganarasimha D, Savelli F, Knierim JJ (2018) Egocentric coding of external items in the lateral entorhinal cortex. Science 362:945–949. https://doi.org/10.1126/science.aau4940

Wernle T, Waaga T, Mørreaunet M, Treves A, Moser M-B, Moser EI (2018) Integration of grid maps in merged environments. Nat Neurosci 21:92–101. https://doi.org/10.1038/s41593-017-0036-6

Whishaw IQ, Hines DJ, Wallace DG (2001) Dead reckoning (path integration) requires the hippocampal formation: evidence from spontaneous exploration and spatial learning tasks in light (allothetic) and dark (idiothetic) tests. Behav Brain Res 127:49–69

Whishaw IQ, Maaswinkel H (1998) Rats with Fimbria–Fornix lesions are impaired in path integration: a role for the hippocampus in "Sense of Direction". J Neurosci 18:3050–3058. https://doi.org/10.1523/JNEUROSCI.18-08-03050.1998

Wiener SI (1993) Spatial and behavioral correlates of striatal neurons in rats performing a self-initiated navigation task. J Neurosci 13:3802–3817. https://doi.org/10.1523/JNEUROSCI.13-09-03802.1993

Wilber AA, Clark BJ, Forster TC, Tatsuno M, McNaughton BL (2014) Interaction of egocentric and world-centered reference frames in the rat posterior parietal cortex. J Neurosci 34:5431–5446. https://doi.org/10.1523/JNEUROSCI.0511-14.2014

Wills TJ, Cacucci F, Burgess N, O'Keefe J (2010) Development of the hippocampal cognitive map in preweanling rats. Science 328:1573–1576. https://doi.org/10.1126/science.1188224

Wilson MA, McNaughton BL (1993) Dynamics of the hippocampal ensemble code for space. Science 261:1055–1058. https://doi.org/10.1126/science.8351520

Wilson MA, McNaughton BL (1994) Reactivation of hippocampal ensemble memories during sleep. Science 265:676–679

Winter SS, Clark BJ, Taube JS (2015a) Disruption of the head direction cell network impairs the parahippocampal grid cell signal. Science 347:870–874. https://doi.org/10.1126/science.1259591

Winter SS, Mehlman ML, Clark BJ, Taube JS (2015b) Passive transport disrupts grid signals in the parahippocampal cortex. Curr Biol 25:2493–2502. https://doi.org/10.1016/j.cub.2015.08.034

Witter MP (1993) Organization of the entorhinal-hippocampal system: a review of current anatomical data. Hippocampus 3 Spec No:33–44

Witter MP, Amaral DG (2004) Chapter 21: Hippocampal formation. In: Paxinos G (ed) The rat nervous system, 3rd edn. Academic Press, Burlington, pp 635–704. https://doi.org/10.1016/B978-012547638-6/50022-5

Witter MP, Room P, Groenewegen HJ, Lohman AHM (1988) Reciprocal connections of the insular and piriform claustrum with limbic cortex: An anatomical study in the cat. Neuroscience 24:519–539. https://doi.org/10.1016/0306-4522(88)90347-8

Wood ER, Dudchenko PA, Robitsek RJ, Eichenbaum H (2000) Hippocampal neurons encode information about different types of memory episodes occurring in the same location. Neuron 27:623–633. https://doi.org/10.1016/S0896-6273(00)00071-4

Yartsev MM, Witter MP, Ulanovsky N (2011) Grid cells without theta oscillations in the entorhinal cortex of bats. Nature 479:103–107. https://doi.org/10.1038/nature10583

Yoganarasimha D, Knierim JJ (2005) Coupling between place cells and head direction cells during relative translations and rotations of distal landmarks. Exp Brain Res 160:344–359. https://doi.org/10.1007/s00221-004-2016-9

Yoganarasimha D, Yu X, Knierim JJ (2006) Head direction cell representations maintain internal coherence during conflicting proximal and distal cue rotations: comparison with hippocampal place cells. J Neurosci 26:622–631. https://doi.org/10.1523/JNEUROSCI.3885-05.2006

Yoo S-W, Lee I (2017) Functional double dissociation within the entorhinal cortex for visual scene-dependent choice behavior. eLife 6:e21543. https://doi.org/10.7554/eLife.21543

Yoon T, Okada J, Jung MW, Kim JJ (2008) Prefrontal cortex and hippocampus subserve different components of working memory in rats. Learn Mem 15:97–105. https://doi.org/10.1101/lm.850808

Young BJ, Fox GD, Eichenbaum H (1994) Correlates of hippocampal complex-spike cell activity in rats performing a nonspatial radial maze task. J Neurosci 14:6553–6563. https://doi.org/10.1523/JNEUROSCI.14-11-06553.1994

Zhang S, Schönfeld F, Wiskott L, Manahan-Vaughan D (2014) Spatial representations of place cells in darkness are supported by path integration and border information. Front Behav Neurosci 8:222. https://doi.org/10.3389/fnbeh.2014.00222

Zhang S-J, Ye J, Miao C, Tsao A, Cerniauskas I, Ledergerber D, Moser M-B, Moser EI (2013) Optogenetic dissection of entorhinal-hippocampal functional connectivity. Science 340:1232627. https://doi.org/10.1126/science.1232627

Zhao R, Grunke SD, Keralapurath MM, Yetman MJ, Lam A, Lee T-C, Sousounis K, Jiang Y, Swing DA, Tessarollo L, Ji D, Jankowsky JL (2016) Impaired recall of positional memory following chemogenetic disruption of place field stability. Cell Rep 16:793–804. https://doi.org/10.1016/j.celrep.2016.06.032

Zingg B, Dong H-W, Tao HW, Zhang LI (2018) Input–output organization of the mouse claustrum. J Comp Neurol 526:2428–2443. https://doi.org/10.1002/cne.24502

Zugaro MB, Berthoz A, Wiener SI (2001) Background, but not foreground, spatial cues are taken as references for head direction responses by rat anterodorsal thalamus neurons. J Neurosci 21:RC154–RC154. https://doi.org/10.1523/JNEUROSCI.21-14-j0001.2001

Neural Circuits for Sleep–Wake Regulation

8

Ying Wu, Lieju Wang, Fen Yang, and Wang Xi

Abstract

The neural mechanisms of sleep, a fundamental biological behavior from invertebrates to humans, have been a long-standing mystery and present an enormous challenge. Gradually, perspectives on the neurobiology of sleep have been more various with the technical innovations over the recent decades, and studies have now identified many specific neural circuits that selectively regulate the initiation and maintenance of wake, rapid eye movement (REM) sleep, and non-REM (NREM) sleep. The cholinergic system in basal forebrain (BF) that fire maximally during waking and REM sleep is one of the key neuromodulation systems related to waking and REM sleep. Here we outline the recent progress of the BF cholinergic system in sleep–wake cycle. The intricate local connectivity and multiple projections to other cortical and subcortical regions of the BF cholinergic system elaborately presented here form a conceptual framework for understanding the coordinating effects with the dissecting regions. This framework also provides evidences regarding the relationships between the general anesthesia and wakefulness/sleep cycle focusing on the neural circuitry of unconsciousness induced by anesthetic drugs.

Keywords

Sleep-wake cycle · Cholinergic neurons · Basal forebrain

Y. Wu · L. Wang
Department of Anesthesiology, The Second Affiliated Hospital, School of Medicine, Zhejiang University, Hangzhou, Zhejiang, China

F. Yang · W. Xi (✉)
Interdisciplinary Institute of Neuroscience and Technology, The Second Affiliated Hospital, School of Medicine, Zhejiang University, Hangzhou, Zhejiang, China
e-mail: XW333@zju.edu.cn

8.1 Introduction

Sleep, which takes up about one-third proportion of mammal life, is a ubiquitous and essential biological need for mammals. Based on diverse behavioral states, electroencephalogram (EEG), and electromyogram (EMG) characteristics, vigilance states can be divided into wakefulness, rapid eye movement (REM) sleep, and non-rapid eye movement (NREM) sleep. The EEG shows low amplitude, fast frequencies, and the EMG shows variable amounts of muscle activity during wake, while EEG is dominated by slower frequencies in the delta (0–4 Hz) and theta (4–7 Hz) ranges during NREM sleep. These three states can be interconvertible between each other and are influenced by environment and physiological signals (including light, ecological niches, temperature, hunger, pain, stress, hormones, metabolic factors, and neurotransmitters) (Brown 2016; Gent et al. 2018).

H. Wang (ed.), *Neural Circuits of Innate Behaviors*, Advances in Experimental Medicine and Biology 1284, https://doi.org/10.1007/978-981-15-7086-5_8

Clinical sleep disorders are closely bound up with general health, including insomnia, narcolepsy (Luppi et al. 2011), paroxysmal activities such as somnambulism, sleep terror disorder, nightmare which occur in the specific sleep period instead of overall time. Short-term sleep deprivation may induce impaired desire to socially interact and feel loneliness (Ben Simon and Walker 2018). Long-term sleep–wake perturbations may lead to chronic emotional, cognitive, and endocrine disorders, including anxiety, depression, Alzheimer's disease (AD), and diabetes (Wang et al. 2015; Cheng et al. 2018; Boeve et al. 2007).

Sleep–wakefulness cycle is a highly complex neural process like a flip-flop pattern (Lu et al. 2006). Dysregulation of neurochemical systems may lead to sleep–awake disorders. It is regulated by various cerebral neurotransmitters and neuromodulators with corresponding specific neuronal ensembles. However, the neural circuits generating sleep–wakefulness cycle and the interaction to other systemic disorders remain a topic of debate. So greater understanding of the neural circuits with different chemical substances of sleep–wakefulness switching is important for further advances in the treatment of illnesses with sleep disorders.

It has been shown that sleep–wake cycle is modulated by a reciprocal interaction between the brainstem and forebrain arousal systems like noradrenergic neurons in the locus coeruleus (LC) or hypocretin neurons in the lateral hypothalamus (LH), and sleep-promoting systems like GABAergic or galanin neurons in the ventrolateral preoptic area (VLPO) (Hobson et al. 1975; Pace-Schott and Hobson 2002). Here we review one of the most popular central arousal neural systems that regulate the sleep/wakefulness cycle: the cholinergic system. The BF and laterodorsal and pedunculopontine tegmental nuclei (LDT/PPT) are the main regions containing acetylcholine, and a 90% proportion of cortical acetylcholine comes from BF. The mazy BF local connectivity, diversity and reciprocity connection pattern from different subcortical systems of the BF cholinergic system, and this cholinergic system to the cortical mantle together forming a neural network modulate arousal behavior (Fig. 8.1). The semblable behavior and EEG characteristics in NREM sleep and anesthesia suggest sleep/awake cycle and anesthesia may share similar neural pathway. The neural network revolving around BF cholinergic neurons can provide evidences and direction about the prospective mainstream perception that the anesthetic-induced loss of consciousness may be due to acting upon the wakefulness/sleep-regulated neural circuits.

8.2 BF Subregions and Cholinergic Neurons Functions in Wakefulness Regulation

A wide variety of studies over the last decades have substantiated that the BF neurons acting as a relay station with afferents from multiple brainstem regions relay to the cortical and subcortical limbic regions. A heterogeneous population of neurons in BF, with the intermingled 5% cholinergic, 55% glutamatergic, 35% GABAergic neurons (Gritti et al. 2006; Henny and Jones 2006, 2008), is topographically scattered in extended forebrain territories across several different cytoarchitectonic areas. BF can be anatomically divided into vertical diagonal band nucleus (VDB), medial septal nucleus (MS), horizontal diagonal band nucleus (HDB), magnocellular preoptic nucleus (MCPO), substantia innominata (SI), nucleus basalis magnocellularis (NBM) (Semba 2004; Rye et al. 1984). The different subsets of BF cholinergic neuron subregions can cast to the discrete areas of cortical and subcortical regions (Zaborszky and Duque 2000; Zaborszky et al. 1999; Rho et al. 2018). For instance, caudal SI and NBM cholinergic neurons innervating to medial frontal cortex and amygdala are wake/REM active (Poulin et al. 2006). The rostral MS and VDB cholinergic axons to dorsal hippocampus can generate cortical theta oscillations (Dutar et al. 1995; Cobb et al. 1999; Agostinelli et al. 2019; Salib et al. 2019), while HDB and MCPO cholinergic neurons to prefrontal cortex (PFC) and olfactory bulb (OB) can promote neocortical activation and

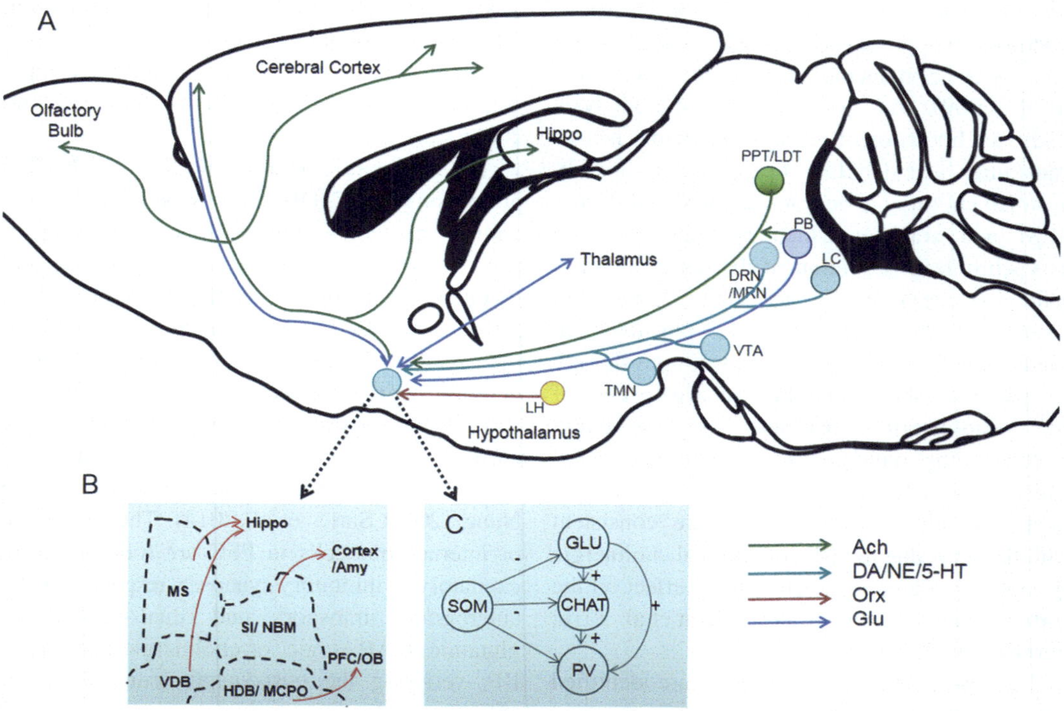

Fig. 8.1 Basal forebrain cholinergic neural circuits diagram for wake–sleep cycles. (**a**) Basal forebrain cholinergic system acts as a relay station which accept various neurochemical sources and own multiple projections to cortical and subcortical regions, forming a general framework. The BF cholinergic projections to cortex and hippocampus (dark green) take the main responsibility for the neocortical activation and hippocampal theta oscillations generation. Meanwhile, the afferents which drive the BF cholinergic neurons mainly originate from the limbic system, diencephalon, brainstem. Monoaminergic neurons (light blue) which are mainly located in caudal hypothalamus and rostral brainstem may directly innervate to the cortex as well as the BF and thalamus. The monoaminergic territory includes LC-NE neurons, DRN and MRN serotonergic neurons, VTA-DA neurons, and TMN histaminergic neurons. The wake-promoting region like LH orexinergic neurons (orange) also cast to BF. PPT/LDT and PB cholinergic neurons (dark green) give rise to highly diffuse projections to BF. Existing glutamatergic (purple) axon terminals in BF arise from multiple afferent sources, including mPFC, vPFC, PB, thalamus, and amygdala. (**b**) The BF cholinergic neurons are topographically segregated within several subsets and may cast to different brain regions, respectively. For example, MS/VDB to hippocampus, SI/NBM to PFC and amygdala, HDB/MCPO to olfactory bulb and PFC. (**c**) The heterogeneous population of neurons in BF is intermingled with 5% cholinergic (ChAT), 55% glutamatergic (Glu), 35% GABAergic neurons, and the GABAergic neurons can be divided into several kinds of subtypes including parvalbumin-positive (PV) and somatostatin (SOM) neurons. In the local microcirculation, SOM neurons inhibit the neighboring wake-promoting Glu, ChAT, and PV neurons, meanwhile, glutamatergic neurons powerfully promote wakefulness through excitation of ChAT and PV neurons directly or indirectly

process sensory (Chaves-Coira et al. 2018; Zheng et al. 2018; Hamamoto et al. 2017).

In spite of the fewer amount of cholinergic neurons in BF, the effect of BF cholinergic neurons in arousal has been subsequently substantiated as a necessity. Researchers have provided a preliminary view of the dynamic interplays between BF-specific cell types and cortical activity via EEG, in vivo electrophysiology, and expressed neurotransmitters (Zaborszky and Duque 2000; Duque et al. 2000). Inconsistent results with different methods or BF subsets function of sleep/awake cycle have been reported. The direct evidence is the expression of c-Fos in the cholinergic neurons during wakefulness, displaying that the 12.9% of BF cholinergic neurons expressed increased c-Fos during spontaneous wakefulness, while 1.8% of cholinergic neurons in spontaneous sleep group (McKenna et al. 2009). As discussed by microdialysis study, the liberation of the slow excitatory neurotransmitter acetylcholine in neocortex, mainly from BF cholinergic neurons, is increased during REM and waking, which is related to the desynchronized activation of the cortex in EEG pattern (Celesia and Jasper 1966; Jasper and Tessier 1971). In normal physiological conditions, in vivo single-unit electrophysiology recording has shown that the discharge pattern of BF cholinergic neurons is positively corresponding with gamma activity and negatively coordinated with delta activity during the sleep/wakefulness cycle, which are consistent with the view that the BF corticopetal cholinergic system exerts a general activational effect on the cortical mantle (Lee et al. 2005; Kim et al. 2016; Berntson et al. 2002).

The rostral MS and VDB in BF are identified important for normal hippocampal activation and theta oscillations generation (Zhang et al. 2011; Kang et al. 2017; Dragoi et al. 1999). Subtypes of nicotinic and muscarinic receptors in the hippocampus are present at presynaptic and postsynaptic location of both principal neurons and inhibitory interneurons, where they exert profound bidirectional influences on synaptic transmission, portraying the key role for cholinergic activation in the induction and maintenance of synaptic plasticity. The cholinergic system is posited as the pacemaker of the hippocampus theta oscillations and undertakes a vital role in vigilance and cognitive function (Drever et al. 2011). Early pharmacological researches have also proved the critical role of acetylcholine in cortical activity, for example, the cholinergic neurons have the capacity to discharge in rhythmic bursts and stimulate cortical gamma and theta activation along with the states of waking and REM while neurotensin (particular agonists of cholinergic neurons) or noradrenaline is administered into the BF (Cape et al. 2000; Jones 2004; Sainsbury and Bland 1981). To the contrary, pharmacological infusion of muscarinic cholinergic antagonist scopolamine or neurotoxic lesions (e.g., 192 IgG-saporin, an immunotoxin selective to cholinergic neurons), into the rostral BF can reduce or abolish septohippocampal neurons' theta rhythmic activity (Leung et al. 2003; Apartis et al. 1998).

The cholinergic neurons in HDB and MCPO projection to PFC, OB, S1 may be specialized in cortical activation and sensory processing of somatosensory stimuli via neuronal tracing and optogenetic manipulations (Chaves-Coira et al. 2016). It has been proved that cortical acetylcholine (Ach), mainly from the BF cholinergic neurons, aggrandize sensory information process and enhance the responsiveness of somatosensory cortical neurons via activation of nicotinic and muscarinic receptors (Martin-Cortecero and Nunez 2014; Sarter et al. 2014). The pyramidal or interneurons cells in PFC are found to form excitatory or inhibitory synapses, respectively, by anterograde transport and dual-stained for glutamic, GABAergic, or cholinergic neurons of BF, verifying the postsynaptic constituents of MCPO cholinergic projection to the PFC (Henny and Jones 2008).

The SI/NBM of BF is deemed to be wake/REM active and may promote neocortical activation. Nonspecific pharmacological adenosine (AD, a modulator of the sleepiness by inhibiting cholinergic and non-cholinergic wakefulness-promoting BF neurons at the AD A1 receptor) and muscimol along with specific immunotoxin 192 IgG-saporin administration into SI/NBM all reduce gamma-EEG power and may promote the transition from wakefulness to NREM. However, other results showed that inactivation or lesion of SI/NBM cholinergic neurons do not alter the sleep–wake profile and reduce the amount of wakefulness, suggesting that BF cholinergic neurons play an enabling role but not necessary in the maintenance of wakefulness (Berntson et al. 2002; Bassant et al. 1995; Wenk et al. 1994; Strecker et al. 2000; Kalinchuk et al. 2008; Kaur et al. 2008; Blanco-Centurion et al. 2006). Thus, the slightly different effect of pharmacologic lesions of BF cholinergic neurons may be accounted for the inassimilable effect of the BF subregions to cortical projection, and the difference release amounts of acetycholine in the various physiological conditions. Further researches confirmed the specific contribution of causal cholinergic SI/NBM neurons in arousal behavior by using genetically targeted optogenetic or chemogenetic manipulations. Selective lesions or inhibition of the SI/NBM cholinergic neurons by optogenetic or chemogenetic manipulations combining with electrophysiology also show the similar consequences to the pharmacological effect (Chen et al. 2016; Fuller et al. 2011). On the contrary, selective optogenetical photoactivation of the SI cholinergic neurons was adequate to disrupt the ongoing sleep state, facilitate an immediate switching to waking or REM sleep from NREM sleep, and prolong the waking durations, providing a direct causal link between cholinergic BF neurons and both cortical activation and arousal behavior (Duque et al. 2000; Chen et al. 2016; Anaclet et al. 2015; Han et al. 2014; Irmak and de Lecea 2014; Hassani et al. 2009). What is interesting among these results is that photostimulation of the cholinergic BF neurons evoked wakefulness only from NREM but not from REM sleep, which present a not quite same as the previous research about the arousal-active cells, for instance, photostimulation of the noradrenergic neurons in the locus coeruleus (LC) or hypocretin neurons in the lateral hypothalamus (LH) induced wake transitions from both NREM and REM sleep (Jones 2008; Adamantidis et al. 2007). In vivo two-photon calcium imaging of neurons in layers 2/3 of mouse visual cortex shows that muscarinic ACh receptors (mAChRs) can activate the excitatory and PV+ neurons in cortex during low levels of cortical desynchronization, and on the contrary, nicotinic ACh receptors (nAChRs) suppressed the excitatory and PV^+ neurons in cortex when cortical desynchronization was strong. Thus, cholinergic input from the BF causes a significant shift in the relative activity levels of different subtypes of cortical neurons through increasing levels of cortical desynchronization (Alitto and Dan 2012).

In addition, the cortical activity influenced by the BF cholinergic inputs can undergo not only directly but also indirectly through the reticular thalamic nuclei, which consists of the relay neurons that may affect the discharge pattern of cortical neurons and can receive the cholinergic afferents at the same time (Jourdain et al. 1989). The projecting BF cholinergic neurons are not restricted to the above well-known regions; they also disseminate to the posterior hypothalamus (Semba and Fibiger 1989). The posterior hypothalamus was firstly put forward as a wakefulness-promoting region because long-playing sleep of encephalitis lethargica may be affiliated with the lesion in posterior hypothalamus and midbrain (Economo 1930). Subsequent studies have also identified the exact effect of cholinergic inputs to posterior hypothalamus that bilateral intracerebral microinjection of muscimol into the posterior hypothalamus induced a transient intensive hypersomnia and can reverse the insomnia induced by preoptic area of hypothalamus (POA) lesion (Sallanon et al. 1989; Lin et al. 1989). In general, up to now, BF cholinergic neurons of each subregions to cortical or subcortical limbic regions form a complex ascending and descending chains regulating sleep/wake cycles (Table 8.1). The diverse inconclusive terminal projected neurons and its synapses from corticopetal BF cholinergic populations make the mechanism of precise modulation of sleep/wakefulness cycles more difficult to a higher degree.

8.3 The Local Microcirculation of BF

The discrepancies between effects of selective and non-selective lesions of the BF cholinergic neurons, as well as the elaborate differential control of electrical stimulations BF on cortical multiunit activity suggested that except for the cholinergic neurons, other types of neurons in BF must become parts of the regulation systems of cortical activity (JimenezCapdeville et al. 1997; Detari 2000). Cholinergic neurons constitute only around 5% fraction in rats BF (Gritti et al. 2006), while various cell populations contain spatially intermingled sleep- and wake-active neurons, including glutamatergic and different subtypes of GABAergic neurons which are heterogeneously distributed across spatially distinct

Table 8.1 The BF cholinergic subregions, its outputs, and the effects of activating or inactivating BF cholinergic neurons to sleep/wake cycles

BF cholinergic subregions	Main cholinergic outputs	Normal physiological conditions	Activation methods	Effect of activation of sleep–wake and EEG	Inactivation or lesion methods	Effect of inactivation or lesion on sleep–wake and EEG
Rostral BF (MS, VDB)	Hippocampus, entorhinal cortex	Generate hippocampal theta oscillations (Zhang et al. 2011; Kang et al. 2017; Dragoi et al. 1999)	• Pharmacological: neurotensin (particular agonists of cholinergic neurons) • Administered noradrenaline(NA) into the basal forebrain in naturally sleeping–waking rats (Cape et al. 2000; Jones 2004)	• Stimulate cortical gamma and theta activity with the states of waking and REM	• Pharmacological muscarinic cholinergic antagonist scopolamine (Leung et al. 2003)	• Reduce or abolish hippocampal theta activity • Scopolamine decreased the LTP induced during walking but did not affect that induced during awake-immobility, did not reduce the LTP during walking
					• Neurotoxic lesions, e.g., 192 IgG-saporin-treated (Sainsbury and Bland 1981; Leung et al. 2003; Apartis et al. 1998)	• Reduce or abolish septohippocampal neurons' theta rhythmic activity • IgG192-saporin showed no difference in the LTP induced during walking and awake-immobility
HDB/ MCPO	Olfactory bulb, PFC	• Wake/REM active, promote neocortical activation • Sensory processing (Chaves-Coira et al. 2016)	• Optogenetic stimulation combining transgenic mice (Chaves-Coira et al. 2016)	• Optogenetic HDB stimulation induced more extensive facilitation of tactile evoked potentials in primary somatosensory than auditory evoked potentials in primary auditory	None	
			• Local microinjection selective serotonin 5-HT_{2C} receptor agonist or 5-HT_6 receptor agonist (Monti et al. 2013; Monti and Jantos 2015)	• Decreased REM sleep without significantly altering wakefulness or SWS		

Caudal BF (SI, NBM)	PFC, amygdala	• Wake/REM active, promote neocortical activation	• Optogenetic or chemogenetic manipulations combining with electrophysiology (Fuller et al. 2011; Anaclet et al. 2015; Han et al. 2014; Hassani et al. 2009)	• Evoked wakefulness only from NREM but not from REM sleep	• Pharmacological adenosine (inhibit cholinergic and non-cholinergic wakefulness-promoting BF neurons at the adenosine A1 receptor) or muscimol inactivation (Strecker et al. 2000; Kalinchuk et al. 2008; Blanco-Centurion et al. 2006)	• Extracellular adenosine accumulates selectively in the BF and cortex and promotes the transition from wakefulness to SWS • Induced slow waves in the neocortex of rats while the animals remained standing and responsive to sensory stimuli
					• Local intraparenchymal infusions of the immunotoxin 192 IgG-saporin (Berntson et al. 2002; Kaur et al. 2008)	• Reduce gamma-EEG power but not significantly alter the overall proportion of sleeping and waking states • Not be essential for sleep/wake stage-switching
				• Induced a rapid desynchronization of the EEG and an increase in EMG power • Elicit cortical activation and facilitate state transitions, particularly transitions to wakefulness and arousal	• Selective lesions or inhibition of the cholinergic neurons by optogenetic or chemogenetic manipulations combining with electrophysiology (Chen et al. 2016; Hassani et al. 2009)	• Enhance slow EEG rhythms and reduce awakening from NREM sleep, but did not alter the sleep–wake profile and not reduce the amount of wakefulness • Chemogenetic inhibition of BF cholinergic neurons increased EEG delta power spectrum and slightly decreased wakefulness
				• Chemogenetic activation of BF cholinergic neurons decreased the EEG delta power spectrum, produced low-delta NREM sleep, and increased wakefulness		

clusters (Nadasdy et al. 2010; Castaneda et al. 2005).

Except for doubtless cholinergic neurons effect on sleep/wakefulness cycle, the GABAergic BF neurons are heterogeneous anatomically and functionally, subdivided on the basis of immunostaining expression of parvalbumin, calretinin, calbindin, somatostatin, Kv2.2, and other markers (Lin et al. 2015; Gritti et al. 2003). The GABAergic neurons can modulate wakefulness apart from NREM sleep beyond expectation, and this effect of GABAergic neurons depends on the cell subtypes and its upstream and downstream neurons. Researchers revealed a critical contribution of the BF GABAergic neurons to drive wake and fast cortical rhythms in behaving mice because chemogenetic activation of the GABAergic neurons reduces the activity of inhibitory cortical interneurons and promote the cortical activation, and these GABAergic neurons have indirect effects via projections to the arousal-related midline thalamus nuclei and many other subcortical arousal regions (Anaclet et al. 2015; Kim et al. 2015). Further studies showed that the parvalbumin-positive (PV+) GABAergic neurons in BF are wake/REM active, and optogenetic activation of this cell type can rapidly induce wakefulness and elicit cortical gamma band oscillations (GBOs, ~40 Hz activity), while photostimulation of the somatostatin (SOM+) neurons may increase NREM sleep mildly (Kim et al. 2015; Xu et al. 2015). The expression of biomarkers released by GABAergic neurons may account for the distinct reaction, and further experiments are needed to clarify the GABAergic subtypes among the sleep/wakefulness regulation.

The glutamatergic neurons account for about 55% proportion in BF (Gritti et al. 2006). Two anatomical pathways of the BF glutamatergic neurons consist of either direct projections to entorhinal cortex and prefrontal cortex or indirect projections to cortex by exciting the local cholinergic neurons with the synaptic connections (Henny and Jones 2008; Manns et al. 2001). Optogenetic activation and recordings from channelrhodopsin-2 (ChR2)-labeled neurons revealed that BF glutamatergic neurons are wake/REM active, and activation of this cell type can rapidly induce wakefulness (Xu et al. 2015). Glutamatergic neuron-specific lesions of BF and its upperstream pontine parabrachial nucleus and adjacent precoeruleus area (PB-PC) complex produced behavioral unresponsiveness like vegetative state, increased EEG delta power, and lack of cortical c-Fos expression during gentle handling, indicating that the PB-PC-BF-cortical pathway may play a critical role in arousal in rats (Fuller et al. 2011; Saito et al. 1977). Then researchers map the local synaptic connections of the main cell types of BF by ultrastructural studies (Zaborszky and Duque 2000), in vitro pharmacology (Yang et al. 2014), and in vivo microdialysis (Zant et al. 2016). The excitatory potential transmit gradationally with the sequence of glutamatergic-cholinergic-PV+ GABAergic neuron connections, and they all received strong inhibition from SOM+ GABAergic neurons via local synapses or long-range projections (Xu et al. 2015; Zant et al. 2016; Weber and Dan 2016). All the above results suggested that noncholinergic BF neurons promote cortical activation by inhibiting delta waves, whereas cholinergic BF neurons play a nonexclusive role in promoting wake, deciphering the basic local organization of the BF circuit for sleep–wake control.

8.4 Afferents to BF Cholinergic Neurons in Mediating Arousal

Anatomical studies in rodents have shown that the afferents which drive the BF cholinergic neurons mainly originate from the limbic system, diencephalon, brainstem (Semba and Fibiger 1989; Zaborszky et al. 1991), including locus coeruleus (LC), ventral tegmental area (VTA) (Trulson and Preussler 1984), median raphe (MRN) (Hajszan and Zaborszky 2002; Smiley et al. 1999; Zaborszky and Luine 1987), dorsal raphenuclei (DRN) (Brown et al. 2002), tuberomammillary nucleus (TMN) (Zant et al. 2012), lateral hypothalamus (LH) (Zaborszky et al. 1993; Zaborszky and Cullinan 1989), basolateral amygdala (Zaborszky et al. 1984) and laterodorsal and pedunculopontine tegmental nuclei (LDT/PPT) in brainstem (Losier and Semba 1993), many of these regions have been certified as wakefulness-promoting nucleuses and form interactional network to promote waking. Nowadays, researchers have even suggested a challenging viewpoint that the brainstem-BF-cortex pathway may play a more important role in arousal regulation than the limited contribution of the traditional reticular formation-thalamus-cortex ascending arousal system, which is necessary for the cognition instead of the arousal state (Fuller et al. 2011).

8.4.1 Monoaminergic Effects via BF Cholinergic Neurons on Arousal

Monoaminergic neurons are mainly known as wake-promoting neurons, including dopamine (DA)-containing VTA neurons (Trulson and Preussler 1984), serotonin-containing DRN neurons (Kirby et al. 2003), NA-containing LC neurons (Aston-Jones and Waterhouse 2016; Schwarz and Luo 2015), or histamine-containing TMN neurons (Haas and Panula 2003). These monoaminergic neurons fire with similar patterns, high-frequency firing rate while the host are awake, slow frequency firing rate while NREM sleep, and shutoff of firing rate during REM sleep (Saper et al. 2010). The BF cholinergic neurons acting as a relay station can take in the multiple monoaminergic inputs to transmission to neocortex in sleep–wake cycle.

NE has long been confirmed a critical modulator of behavioral arousal that is necessary for the maintenance of cortical tonic EEG activation (Jouvet 1972). In another early study, noradrenaline was depleted by a mean of 85% in the paleo- and neocortex after complete bilateral lesions of the NE in cats. However the EEG activation reappear within 12–48 h following the lesion, indicating that the NE-LC neurons are not necessary for the tonic maintenance of EEG activation in wakefulness and in amphetamine-produced arousal (Barbara et al. 1977). Furthermore, it is illustrated by optogenetic tool that there is a causal relationship among LC firing, cortical activity, sleep–wake switch, and locomotor arousal (Carter et al. 2010). The excitatory effect of noradrenalin into SI was blocked by the $alpha_1$-adrenergic receptor antagonist prazosin and not by the alpha 2-antagonist yohimbine, suggesting the direct excitatory pathway (Fort et al. 1995). On the other hand, $alpha_{2A}$-adrenergic receptors were found to be localized in a large amount of non-cholinergic neurons (putative GABAergic neurons) except cholinergic neurons. NE may disinhibit the GABAergic to cholinergic link, thereby causing an indirect facilitatory action on cholinergic neurons. This indirect excitatory effect of NE to BF cholinergic neurons may be paralleled with direct excitatory action. Noradrenaline microinjection to BF in free-moving rats produced a dose-dependent increase in gamma-EEG activity, a decrease in delta activity, and an increase in arousal by depolarizing and exciting the BF cholinergic neurons, suggesting the adrenergic/cholinergic direct link in BF cholinergic neurons may represent a critical component of a central network coordinating autonomic regulation with cortical activation and arousal (Hajszan and Zaborszky 2002; Cape and Jones 1998; Espana and Berridge 2006). Thus, considering the presence of heterogeneous neuronal populations of BF and the various adrenergic receptors, LC-NE may affect cortical activation by BF according to a complicated cellular mechanism.

It is more widely accepted that VTA-DA neurons drive the motivated behaviors (Bromberg-Martin et al. 2010). The first study indicating the possible capacity to arousal of the DA is that DA could also be implicated in the mechanism of action of neuroleptic drugs (Carlsson and Lindqvist 1963). Then the dopamine-containing neurons located in the VTA and the substantia nigra pars compacta (SNc) were proposed to involve in the maintenance of behavioral arousal (Trulson and Preussler 1984). Although the discharge pattern of VTA-DA neurons and SNc are not significantly regulated with the sleep/wake cycle or anesthetics administration (Miller et al. 1983; Steinfels et al. 1983), administration of DA-D1-receptor agonists SKF 38393 produces dose-dependently EEG desynchronization related to behavioral arousal. Deletion of the dopamine transporter (DAT) gene in mice reduced NREM sleep, increased wakefulness consolidation, and moderately increased wheel-running activity only during the latter portion of the dark period, indicating that the VTA-DA neurons are wakefulness-promoting (Ongini et al. 1985; Wisor et al. 2001). The SNc/VTA-DA-containing neurons have bidirectional interaction with the other monoaminetic neurons, cholinergic PPT/LDT neurons, Ach- and PV-containing BF neurons, orexin-containing LH neurons that are dominant cells in the process of sleep–wake transition (Eban-Rothschild et al. 2016; Moore and Bloom 1978). The cholinergic neurons in slices of BF were carried via whole-cell patch-clamp recordings with the non-N-methyl-D-aspartate (NMDA) glutamatergic excitatory postsynaptic currents (EPSCs), illuminating D1-like receptor-mediated presynaptic inhibition of glutamate release onto cholinergic BF neurons (Momiyama 2010; Momiyama and Nishijo 2017). However, it is interesting to note that VTA non-dopaminergic (putative somatostatin-containing) neurons increase firing rates during active wakefulness and REM sleep and project to the GABAergic SI and adjacent BF innervate cholinergic corticopetal areas, forming VTA-GABAergic/BF-GABA-ACh/cortex pathway that may be another arousal-related route (Zaborszky 1989; Marazioti et al. 2005; Smith et al. 2001).

Most forebrain and cortical serotonin emanate from the DRN and MRN. Early studies have identified that the early electrophysiological properties of serotoninergic neurons in DRN showing state-dependent changes during wake/sleep cycles, like other wake-regulating monoaminergic cells (Kirby et al. 2003; Celada et al. 2013; McGinty and Harper 1976). Serotonin brings about cortical activation directly by enhancing the membrane excitability of neocortical pyramidal neurons, while serotonin-dependent cortical activation is not disturbed by cholinergic, dopaminergic, noradrenergic, or histaminergic blockade, suggesting that the excitatory effect of serotonin release is present by its direct active impact to the cortex (Monti 2010; Servos et al. 1994; Vanderwolf and Baker 1986). Early studies showed that cholinergic neurons and calretinin-containing GABAergic neurons in MS have no serotoninergic fibers synapses innervation, but PV-positive neurons in MS have serotoninergic fibers synapses innervation (Leranth and Vertes 1999). Neuroanatomical evidence have further shown that serotonergic terminals were identified in immediate connection with the cholinergic neurons in a subpopulation of SI, MS, VDB, and NBM (Kia et al. 1996; Muzerelle et al. 2016), and the presynaptic inhibition of the GABA release onto cholinergic neurons is mediated by 5-HT1B receptors (Momiyama and Nishijo 2017). Local microinjection of selective serotonin 5-HT2C receptor agonist or 5-HT6 receptor agonist into HDB produces a dose-dependent decrease in gamma-EEG activity and REM sleep without significantly altering amounts of wakefulness or SWS, suggesting that the effect of serotoninergic system could be partly related to the Ach release of cholinergic neurons in the frontal cortex and hippocampus (Cape and Jones 1998; Monti et al. 2013; Monti and Jantos 2015; Khateb et al. 1993). Thus the BF might be a relay station conveying neural information from serotoninergic cells toward cortex.

The histaminergic neurons exist only in a posterior hypothalamic region, TMN, spreading widely to cortex, thalamus, and other arousal-promoting regions and synergistically promote cortical activation and wake (Zant et al. 2012; Haas and Panula 2003; Yu et al. 2018; Thakkar 2011). Manipulations of extracellular histamine concentrations in TMN and BF have consistent link to behavioral state. Increased histaminergic transmission increases the state of wakefulness, while lessened histamine concentrations increase NREM sleep. Lesioning the BF cholinergic neurons abolished these links between levels of BF and behavioral state (Thakkar 2011; Lin et al. 2011; Ramesh et al. 2004). Microinjection of H1 but not H2 or H3 receptor agonists to the NBM imitated the effect of histamine into BF and altered ACh spontaneous release from the cortex, demonstrating that activation of histamine H1 receptors in the NBM induced cortical activation. On the contrary, the reductive wakefulness and cortical activation were shown after histamine receptor 1 antagonist perfusion (Cecchi et al. 2001). These experiments clarified that the histaminergic afferent to BF cholinergic neurons are wakefulness-promoting, but the histaminergic afferent projecting subsets of BF is still arcane.

8.4.2 Lateral Hypothalamus Orexinergic Inputs to the BF Cholinergic Neurons

The LH orexinergic neurons are wakefulness-promoting system, while lack of orexin neurotransmission produces a chronic state of hypoarousal characterized by excessive sleepiness named narcolepsy, frequent transitions between wake and sleep, and episodes of cataplexy (Thannickal et al. 2000; Mochizuki et al. 2011). Pharmacological infusion of orexin-A or photoactivation or chemogenetic activation of the orexinergic neurons can increase the probability of transition to wakefulness from either SWS or REM sleep by modulating the arousal threshold (Adamantidis et al. 2007; Sutcliffe and de Lecea 2002). Thus, these studies are in favor of the significance between the orexin neurons activity and mammalian wakefulness state. Anatomical studies have shown that these orexinergic neurons project to diverse arousal behavioral-related regions, such as the BF, TMN, DR, LC, pontine reticular formation, and the PPT/LDT nuclei, in addition to the neocortex (Nambu et al. 1999; Peyron et al. 1998). Orexinergic fibers synapses have been found on BF cholinergic neurons, and orexin-A is released in the BF during waking. Local microdialysis infusion of orexins excites BF cholinergic neurons, induces cortical release of acetylcholine, and strongly promotes wakefulness for several hours (Arrigoni et al. 2010; Fadel et al. 2005; Thakkar et al. 2001). Several in vitro slice recording studies have also shed light on how the orexinergic neurons activate BF, such as MCPO, SI, and MS cholinergic neurons. Orexin-A increases evoked excitatory postsynaptic currents in cholinergic and non-cholinergic (putative GABAergic) corticopetal neurons (Eggermann et al. 2001; Wu et al. 2004). Meanwhile, the orexinergic neurons may also co-release the inhibitory neuropeptide dynorphin, which have inhibitive effects on the sleep-active GABAergic neurons of BF, suggesting orexins and dynorphin may act synergistically in the BF to promote arousal (Gerashchenko et al. 2001). This evidence suggests that the BF is a key relay site through which orexins activate the cortex and promotes behavioral arousal.

8.4.3 LDT/PPT Afferents to the BF Cholinergic Neurons

Since precursory study exhibit that the anesthetic cat switch to the awakening-like EEG after electrical stimulation of brainstem reticular formation, the reticular formation has been proposed to play a critical role in the initiation and maintenance of cortical activation during wakefulness and REM sleep state (Moruzzi and Magoun 1949). Apart from the cholinergic neurons that are scattered across a number of classically defined subregions of BF, the other regions in brainstem including medullary reticular formation, LDT/PPT are also rich of cholinergic neurons (Hallanger et al. 1987; Jones et al. 1986). The cholinergic neurons in LDT/PPT discharge the highest during active wake and REM sleep in positive correlation with fast cortical activity while the low discharge in SWS, which are also consistent with the slow cortical activity, as “Wake/REM-max active neurons” (Boucetta et al. 2014). Electrical stimulation of the PPT during sleep leads to rapid awakening, activation of the PPT cholinergic cells by using chemogenetic or optogenetic tools inhibits the slow wave activity during NREM sleep and seizures (Van Dort et al. 2015; Kroeger et al. 2017). Lesions or injection of cholinergic agonist into the PPT/LDT of cats or rats triggers a largely long-term REM with muscle atonia (Shiromani et al. 1996; Baghdoyan et al. 1987;

Webster and Jones 1988). Together, these studies strongly suggest a role of cholinergic cells in PPT/LDT in REM sleep control. However, whether the PPT promotes wake itself still remains unclear. The innervations of PPT/LDT neurons give rise to highly diffuse projections to many arousal-promoting brain regions, including VTA, LH, BF, frontal cortex, and many thalamic nuclei, but the fibers are sparse (Hallanger and Wainer 1988). Only 8% PPT/LDT cholinergic neurons become the BF afferents; however, the detailed function of this pathway is still perplexing (Losier and Semba 1993). The retrograde transport of horseradish peroxidase-conjugated wheatgerm agglutinin (WAG) with immunohistochemistry has shown the other afferents to BF cholinergic neurons from PPT monoamine neurons, including catecholamine, serotonin, and acetylcholine neurons (Jones and Cuello 1989). Single electrical pulses delivered to the PPT area can produce excitatory effects in the majority (72%) of BF fast cortical EEG waves cells (F-cells) mediating the ascending excitatory drive from the brainstem to the cerebral cortex (Detari et al. 1997). Subsequent retrograde trace and pharmacological studies unveiling a disagreement that cortical ACh release and EEG activation evoked by PPT stimulation were not blocked by either cholinergic muscarinic or nicotinic antagonists applied to NBM , but were apparently reduced by glutamate antagonists, suggesting that the excitatory input to BF cholinergic neurons from PPT may be via glutamatergic axons (Rasmusson et al. 1994; Steriade 1995). Novel studies also began to emphasize on the non-cholinergic neurons of the PPT/LDT, such as the glutamatergic and GABAergic neurons, which may have distinct influence on cortical activity and sleep/wake behavior (Boucetta et al. 2014; Kroeger et al. 2017). In general, the connections from PPT/LDT to BF may contain the cholinergic, monoamine and glutamatergic pathway in regulating sleep–wake cycle.

8.4.4 Glutamatergic Afferents to the BF Cholinergic Neurons in Mediating Arousal

Interaction of glutamate and ACh in the BF cholinergic neurons has long been studied in vitro and in vivo. The morphological evidence of the presence of monosynaptic glutamate inputs to BF cholinergic neurons in SI and the glutamatergic receptors lying in the cholinergic neurons suggests that the glutamate inputs affect cholinergic corticopetal neurons (Sim and Griffith 1996; Hur et al. 2009). Moreover, the release of ACh in the cortex and increased gamma and theta EEG activity can be induced by local administration of glutamate agonists kainic acid and N-methyl-d-aspartic acid (NMDA) into the BF, suggesting that glutamatergic inputs can excite BF cholinergic neurons (Cape and Jones 2000; Fournier et al. 2004; Fadel et al. 2001). The existing glutamatergic axon terminals in BF arise from multiple afferent sources, including mPFC, ventral prefrontal cortex (vPFC) (Zaborszky et al. 1997), amygdala (Zaborszky et al. 1984), thalamus, hypothalamus (Carnes et al. 1990), and brainstem reticulate formation (Jones and Cuello 1989; Lavoie and Parent 1994; Semba et al. 1988). The glutamatergic inputs from other regions along with the local BF excitatory glutamatergic neurons make the glutamatergic inputs resources to cholinergic neurons more diversification and make the network more accordant.

For instance, anatomical evidence show that projections from the amygdala terminate in the SI. Systemic administration of centrally acting muscarinic receptor antagonists could block the cholinergically mediated neocortical arousal by unilateral electrical stimulation of the amygdaloid central nucleus (ACe). These results suggested the involvement of the BF in amygdala-induced arousal (McDonald 1991; Kapp et al. 1994; Dringenberg and Vanderwolf 1996). However, microscopic studies showed that paradoxical results, the existence of glutamatergic inputs and the inhibitory GABAergic inputs from the amygdala to BF (Zaborszky et al. 1984; Pare and Smith 1994), are in line with an early study that there were antergic glutamatergic and GABAergic systems in the basolateral amygdala that one desynchronizes while another one synchronizes the neocortex activity (Kreindler and Steriade 1964).

The posterior hypothalamic supramammillary (SuM) nucleus has been considered as a key relay station in the brainstem reticular nucleus pontis oralis (PnO)-hypothalamic SuM-septo-hippocampal ascending pathway, and it has been thought to be a key node in the generation of hippocampal theta rhythm, arousal system, and active movements (Vertes and Kocsis 1997; Vertes 2015; Renouard et al. 2015; Pedersen et al. 2017). Electrical stimulation of the posterior hypothalamic nucleus or its upstream reticular nucleus pontis oralis (RPO) induced cortical activation and hippocampal theta, and the RPO-elicited discharge patterns of all theta-ON cells can be blocked by procaine hydrochloride administration into the MS/VDB, indicating movement-related ascending activation of a hypothalamo-septal pathway (Oddie et al. 1996; Kirk et al. 1996). Early dissected studies by electron microscopic double-immunostaining experiments showed that the glutamatergic fibers of SuM calretinin neurons form synaptic contacts and terminate on both PV-containing GABAergic and cholinergic neurons in the MS/VDB complex (Leranth and Kiss 1996; Borhegyi et al. 1998). Novel results also suggest that tonic activation of limbic cortical neurons during REM sleep is due to projections from GABA/glutamate co-releasing neurons of SuM (Luppi et al. 2017), suggesting that the circuitry from SuM to BF is one of the pathway of regulating sleep/awake cycle. Thus, whether the excitatory or inhibited inputs to BF heterogeneous subsets have different effects to sleep modulation awaits for deeper investigation.

8.5 Interlink BF Cholinergic Systems Between Sleep–Wake Cycle and General Anesthesia

General anesthesia is a man-made neurophysiological state comprising of unconsciousness, amnesia, analgesia, and immobility, making it possible to perform surgery of the patients. The expected clinical general anesthetic drugs should satisfy the desirable outcomes, such as surgery without pain, awareness, and memory, and minimize the undesirable adverse reaction, such as cardiovascular depression, delirium, and even death. To further study the neural mechanism of general anesthetic, the neural mechanism of sleep state is often put toward to imitate the anesthetic process because they both exist invertible loss of consciousness (LOC) and a lack of response to environmental stimuli (Brown et al. 2010). The recording EEG characters also showed depressed neuronal unit activity between NREM sleep and anesthetized animals, including spindles, high-amplitude cortical slow waves of 0–4 Hz (delta oscillations), and reduced muscle tone (Akeju and Brown 2017; Massimini et al. 2005). What is more, functional brain imaging studies regarding propofol and sevoflurane administration showed deactivation of the thalamus and brainstem regions that are identified as the ascending reticular arousal system (Kaisti et al. 2003; Bonhomme et al. 2001). However, whether the two states share the same precise neural circuits remains mysterious, while deciphering the interlink of the two state is important for both neuroscience and pharmaceutical research to facilitate the development of new anesthetics. Among the multitudinous effective brain regions, BF, a relay station, especially the cortical Ach original location, has been delineated an important regulator of the state of consciousness during general anesthesia.

The level of Ach in neocortex is declining during isoflurane general anesthesia and NREM sleep, while release of Ach is in line with the high-frequency cortical activity during waking and REM sleep (Phillis 2005; Dong et al. 2006). In clinical anesthetic operation patients, systemic venous administration cholinesterase inhibitor physostigmine increased central Ach levels and promoted arousal via muscarinic cholinergic receptors during halothane or propofol anesthesia. In early rodents studied, debasement of BF Ach by interventricular administration hemicholinium-3 lower the MAC (minimal alveolar concentration, an index that reflexes the efficiency of inhaled anesthetic), on the contrast, systematic administration physostigmine increased the isoflurane MAC, and cholinergic agents can reverse the electroencephalogram-depressant effect of isoflurane (Kenny et al. 2016; Hudetz et al. 2003; Zucker 1991). However, it is controversial about the effect of physostigmine while it decreased halothane MAC in dogs and was ineffectual in bispectral index (BIS) and in scores assessing early recovery in the emergence of sevoflurane anesthesia patients (Paraskeva et al. 2002). Thus, the conflicting effects of systemic pharmacological cholinergic agonist or antagonist are too extensive to ensure the precise anesthetic target in the brain.

Subsequent researches further overcame the disadvantages of systemic administration of cholinergic agonist or antagonist and highlighted the unique effect of BF and even its cholinergic neurons in regulating anesthesia-induced loss of consciousness. Rats with electrolytic lesion of the MS (mainly projects to the hippocampus) showed increased sensitivity with the reduced ED50 for LORR and delayed the emergence from halothane, isoflurane, pentobarbital, and propofol anesthesia (Leung et al. 2013). Furthermore, to identify the specificity of subtypes of cell in anesthetic-induced unconsciousness, selective 192 IgG-saporin lesion of MS/VDB cholinergic neurons showed increased anesthesia sensitivity to cumulative doses of intraperitoneal injection of propofol with the ED50 for LORR leftward shifting and more decreased hippocampal gamma power during isoflurane anesthesia (Laalou et al. 2008; Tai et al. 2014).

Similarly, lesion by bilateral infusion of 192IgG-saporin into another BF subsets, NBM, showed significantly longer duration of LORR after propofol, pentobarbital but not after halothane (2%) compared to control. Meanwhile, reversible inactivation of NBM with local administration of GABAA receptor agonist muscimol can increase slow waves in neocortex during awake state, and prolonged the duration of LORR and loss of tail-pinch response during propofol, pentobarbital, and halothane. Thus, lesion of NBM cholinergic neurons, playing similar effect like inactivation of the NBM, prolonged the LORR response to general anesthetic drugs (Leung et al. 2011).

Different inputted exogenous neurotransmitter to BF also modulated the anesthetic-induced unconsciousness behavior and sensitivity. Injection of orexin-A and orexin-B into the NBM significantly increased the acetylcholine efflux in the somatosensory cortex, reduced the depth of 1.2% isoflurane anesthesia (1 MAC) as indicated by burst-suppression ratio of the recording epidural EEG and shortened the emergence time. The infusion of orexin-A receptor antagonist (SB-334867A) into NBM delayed the emergence time to sevoflurane and propofol, illuminating the orexinergic afferent into BF may also impact the efficiency of anesthetic (Dong et al. 2009; Zhang et al. 2012). Compared with orexin-B, NBM microinjection of orexin-A was more potent in producing greater Ach efflux in the cortex and greater relief from the burst suppression induced by isoflurane (Zhang et al. 2012). In another study, microinjection of orexin-A into BF facilitated the emergence of rats from isoflurane anesthesia, while orexin-B did not, indicating that orexin-A into BF plays a promotive role in the emergence of isoflurane anesthesia (Zhang et al. 2016).

Parallel results are also shown that histamine and norepinephrine microinfusion into NBM hastened emergence time (recovery of righting) from isoflurane or desflurane anesthesia and more spontaneous movements and frontal EEG desynchronization (Pillay et al. 2011; Luo and Leung 2009).

But whether the above effective postsynaptic neurons in BF are cholinergic neurons is still sciolistic because not only the cholinergic neurons have the orexinergic or histaminergic receptors and the interplay of matched heterogeneous neurons types of BF is intricate. As an example, previous studies have demonstrated the electrophysiological properties of cholinergic and noncholinergic neurons of BF in whole-cell patch-clamp recordings, the ChAT+ neurons charged with lower frequencies than the ChAT– neurons stimulated by tonic depolarization (Lopez-Hernandez et al. 2017; Unal et al. 2012). According to the electrophysiological properties, propofol decreased the excitability of cholinergic neurons in mouse BF via combing with GABAA receptors (Chen et al. 2018). However, studies that clarify the effect to cholinergic neurons by accurate receptors that anesthetic acts on are needed.

Moreover, there are still exceptional paradoxical reports about the probable effect that selective lesion by intracerebroventricular administration of 192 IgG-Saporin into cholinergic neurons in BF alleviated the sedative potency of subanesthetic low-dose propofol (30 mg/kg i.p.) but elevated the anesthesia potency of the anesthetic high-dose propofol (>100 mg/kg i.p.), in which the difference between subanesthetic low dose and anesthetic high dose is still hard to explain (Laalou et al. 2008; Pain et al. 2000).

Finally, the heterogeneity and complex circuits of the BF neurons during sleep–wake cycle determined the general effect of increased NREM sleep by administration of anesthetic is not so simple. How the BF local neurons microcircuitry and other wake/sleep-active nucleus interplay with each other and form the network and its function in anesthesia still have great study space.

In this review, the BF cholinergic sleep/wakefulness circuits are portrayed in a relatively simple pattern; however, the networks among the plentiful neural systems may be much more complicated, flexible, and may be affected by many factors, such as light, temperature, stress, hormones, and so on. New burgeoning technology like genetical tools, optogenetic, calcium imaging, and delicate computational analysis will be required to elucidate the networks. Only with deeper comprehension of the anatomy, physiology, and dynamics of the wake/sleep circuits, we can treat the sleep disorders and the sleep-related cognitive and emotional diseases with the more accurate targets in brain and optimize the clinical general anesthetic drugs to a desirable outcome.

References

Adamantidis AR, Zhang F, Aravanis AM, Deisseroth K, de Lecea L (2007) Neural substrates of awakening probed with optogenetic control of hypocretin neurons. Nature 450:420–424

Agostinelli LJ, Geerling JC, Scammell TE (2019) Basal forebrain subcortical projections. Brain Struct Funct 224(3):1097–1117

Akeju O, Brown EN (2017) Neural oscillations demonstrate that general anesthesia and sedative states are neurophysiologically distinct from sleep. Curr Opin Neurobiol 44:178–185

Alitto HJ, Dan Y (2012) Cell-type-specific modulation of neocortical activity by basal forebrain input. Front Syst Neurosci 6:79

Anaclet C, Pedersen NP, Ferrari LL, Venner A, Bass CE, Arrigoni E, Fuller PM (2015) Basal forebrain control of wakefulness and cortical rhythms. Nat Commun 6:8744

Apartis E, Poindessous-Jazat FR, Lamour YA, Bassant MH (1998) Loss of rhythmically bursting neurons in rat medial septum following selective lesion of septohippocampal cholinergic system. J Neurophysiol 79:1633–1642

Arrigoni E, Mochizuki T, Scammell TE (2010) Activation of the basal forebrain by the orexin/hypocretin neurones. Acta Physiol 198:223–235

Aston-Jones G, Waterhouse B (2016) Locus coeruleus: from global projection system to adaptive regulation of behavior. Brain Res 1645:75–78

Baghdoyan HA, Rodrigoangulo ML, Mccarley RW, Hobson JA (1987) A neuroanatomical gradient in the pontine tegmentum for the cholinoceptive induction of desynchronized sleep signs. Brain Res 414:245–261

Barbara EJ, Stanley TH, Halaris AE (1977) Effects of locus coeruleus lesions upon cerebral monoamine content, sleep-wakefulness states and the response to amphetamine in the cat. Brain Res 124:473–496

Bassant MH, Apartis E, Jazatpoindessous FR, Wiley RG, Lamour YA (1995) selective immunolesion of the Basal forebrain cholinergic neurons—effects on hippocampal activity during sleep and wakefulness in the rat. Neurodegeneration 4:61–70

Ben Simon E, Walker MP (2018) Sleep loss causes social withdrawal and loneliness. Nat Commun 9:3146

Berntson GG, Shafi R, Sarter M (2002) Specific contributions of the basal forebrain corticopetal cholinergic system to electroencephalographic activity and sleep/waking behaviour. Eur J Neurosci 16:2453–2461

Blanco-Centurion C, Xu M, Murillo-Rodriguez E, Gerashchenko D, Shiromani AM, Salin-Pascual RJ,

Hof PR, Shiromani PJ (2006) Adenosine and sleep homeostasis in the basal forebrain. J Neurosci 26:8092–8100

Boeve BF, Silber MH, Saper CB, Ferman TJ, Dickson DW, Parisi JE, Benarroch EE, Ahlskog JE, Smith GE, Caselli RC, Tippman-Peikert M, Olson EJ, Lin SC, Young T, Wszolek Z, Schenck CH, Mahowald MW, Castillo PR, Del Tredici K, Braak H (2007) Pathophysiology of REM sleep behaviour disorder and relevance to neurodegenerative disease. Brain 130:2770–2788

Bonhomme V, Fiset P, Meuret P, Backman S, Plourde G, Paus T, Bushnell MC, Evans AC (2001) Propofol anesthesia and cerebral blood flow changes elicited by vibrotactile stimulation: A positron emission tomography study. J Neurophysiol 85:1299–1308

Borhegyi Z, Magloczky Z, Acsady L, Freund TF (1998) The supramammillary nucleus innervates cholinergic and GABAergic neurons in the medial septum-diagonal band of Broca complex. Neuroscience 82:1053–1065

Boucetta S, Cisse Y, Mainville L, Morales M, Jones BE (2014) Discharge profiles across the sleep-waking cycle of identified cholinergic, GABAergic, and glutamatergic neurons in the pontomesencephalic tegmentum of the rat. J Neurosci 34:4708–4727

Bromberg-Martin ES, Matsumoto M, Hikosaka O (2010) Dopamine in motivational control: rewarding, aversive, and alerting. Neuron 68:815–834

Brown EN, Lydic R, Schiff ND (2010) General anesthesia, sleep, and coma. New Engl J Med 363:2638–2650

Brown RE, Sergeeva OA, Eriksson KS, Haas HL (2002) Convergent excitation of dorsal raphe serotonin neurons by multiple arousal systems (orexin/hypocretin, histamine and noradrenaline). J Neurosci 22:8850–8859

Brown SA (2016) Circadian Metabolism: from mechanisms to metabolomics and medicine. Trends Endocrin Met 27:415–426

Cape EG, Jones BE (1998) Differential modulation of high-frequency gamma-electroencephalogram activity and sleep-wake state by noradrenaline and serotonin microinjections into the region of cholinergic basalis neurons. J Neurosci 18:2653–2666

Cape EG, Jones BE (2000) Effects of glutamate agonist versus procaine microinjections into the basal forebrain cholinergic cell area upon gamma and theta EEG activity and sleep-wake state. Eur J Neurosci 12:2166

Cape EG, Manns ID, Alonso A, Beaudet A, Jones BE (2000) Neurotensin-induced bursting of cholinergic basal forebrain neurons promotes gamma and theta cortical activity together with waking and paradoxical sleep. J Neurosci 20:8452–8461

Carlsson A, Lindqvist M (1963) Effect of chlorpromazine or haloperidol on formation of 3-methoxytyramine and normetanephrine in mouse brain. Acta Pharmacol Tox 20:140

Carnes KM, Fuller TA, Price JL (1990) Sources of presumptive glutamatergic/aspartatergic afferents to the magnocellular basal forebrain in the rat. J Comp Neurol 302:824–852

Carter ME, Yizhar O, Chikahisa S, Nguyen H, Adamantidis A, Nishino S, Deisseroth K, de Lecea L (2010) Tuning arousal with optogenetic modulation of locus coeruleus neurons. Nat Neurosci 13:1526–1533

Castaneda MT, Sanabria ER, Hernandez S, Ayala A, Reyna TA, Wu JY, Colom LV (2005) Glutamic acid decarboxylase isoforms are differentially distributed in the septal region of the rat. Neurosci Res 52:107–119

Cecchi M, Passani MB, Bacciottini L, Mannaioni PF, Blandina P (2001) Cortical acetylcholine release elicited by stimulation of histamine H1 receptors in the nucleus basalis magnocellularis: a dual-probe microdialysis study in the freely moving rat. Eur J Neurosci 13:68–78

Celada P, Puig MV, Artigas F (2013) Serotonin modulation of cortical neurons and networks. Front Integrative Neurosci 7:25

Celesia GG, Jasper HH (1966) Acetylcholine released from cerebral cortex in relation to state of activation. Neurology 16:1053–1063

Chaves-Coira I, Barros-Zulaica N, Rodrigo-Angulo M, Nunez A (2016) Modulation of specific sensory cortical areas by segregated Basal forebrain cholinergic neurons demonstrated by neuronal tracing and optogenetic stimulation in mice. Front Neural Circuit 10:28

Chaves-Coira I, Martin-Cortecero J, Nunez A, Rodrigo-Angulo ML (2018) Basal Forebrain nuclei display distinct projecting pathways and functional circuits to sensory primary and prefrontal cortices in the rat. Front Neuroanat 12:69

Chen, L., Yang, Z. L., Cheng, J., Zhang, P. P., Zhang, L. S., Liu, X. S., Wang, L. C. (2018) Propofol decreases the excitability of cholinergic neurons in mouse basal forebrain via GABAA receptors. Acta Pharmacol Sin 40(6):755-761

Chen L, Yin D, Wang TX, Guo W, Dong H, Xu Q, Luo YJ, Cherasse Y, Lazarus M, Qiu ZL, Lu J, Qu WM, Huang ZL (2016) Basal forebrain cholinergic neurons primarily contribute to inhibition of electroencephalogram delta activity; rather than inducing behavioral wakefulness in mice. Neuropsychopharmacology 41:2133–2146

Cheng W, Rolls ET, Ruan HT, Feng JF (2018) Functional connectivities in the brain that mediate the association between depressive problems and sleep quality. JAMA Psychiat 75:1052–1061

Cobb SR, Bulters DO, Suchak S, Riedel G, Morris RGM, Davies CH (1999) Activation of nicotinic acetylcholine receptors patterns network activity in the rodent hippocampus. J Physiol-London 518:131–140

Detari L (2000) Tonic and phasic influence of basal forebrain unit activity on the cortical EEG. Behav Brain Res 115:159–170

Detari L, Semba K, Rasmusson DD (1997) Responses of cortical EEG-related basal forebrain neurons to brainstem and sensory stimulation in urethane-anaesthetized rats. Eur J Neurosci 9:1153–1161

Dong H, Niu J, Su B, Zhu Z, Lv Y, Li Y, Xiong L (2009) Activation of orexin signal in basal forebrain facilitates the emergence from sevoflurane anesthesia in rat. Neuropeptides 43:179–185

Dong HI, Fukuda S, Murata E, Zhu ZH, Higuchi T (2006) Orexins increase cortical acetylcholine release and electroencephalographic activation through orexin-1 receptor in the rat basal forebrain during isoflurane anesthesia. Anesthesiology 104:1023–1032

Dragoi G, Carpi D, Recce M, Csicsvari J, Buzsaki G (1999) Interactions between hippocampus and medial septum during sharp waves and theta oscillation in the behaving rat. J Neurosci 19:6191–6199

Drever BD, Riedel G, Platt B (2011) The cholinergic system and hippocampal plasticity. Behav Brain Res 221:505–514

Dringenberg HC, Vanderwolf CH (1996) Cholinergic activation of the electrocorticogram: an amygdaloid activating system. Exp Brain Res 108:285–296

Duque A, Balatoni B, Detari L, Zaborszky L (2000) EEG correlation of the discharge properties of identified neurons in the basal forebrain. J Neurophysiol 84:1627–1635

Dutar P, Bassant MH, Senut MC, Lamour Y (1995) The septohippocampal pathway—structure and function of a central cholinergic system. Physiol Rev 75:393–427

Eban-Rothschild A, Rothschild G, Giardino WJ, Jones JR, de Lecea L (2016) VTA dopaminergic neurons regulate ethologically relevant sleep-wake behaviors. Nat Neurosci 19:1356

Economo CV (1930) Sleep as a problem of localization. J Nerv Ment Dis 71:249–259

Eggermann E, Serafin M, Bayer L, Machard D, Saint-Mleux B, Jones BE, Muhlethaler M (2001) Orexins/hypocretins excite basal forebrain cholinergic neurones. Neuroscience 108:177–181

Espana RA, Berridge CW (2006) Organization of noradrenergic efferents to arousal-related basal forebrain structures. J Comp Neurol 496:668–683

Fadel J, Pasumarthi R, Reznikov LR (2005) Stimulation of cortical acetylcholine release by orexin A. Neuroscience 130:541–547

Fadel J, Sarter M, Bruno JP (2001) Basal forebrain glutamatergic modulation of cortical acetylcholine release. Synapse 39:201–212

Fort P, Khateb A, Pegna A, Muhlethaler M, Jones BE (1995) Noradrenergic modulation of cholinergic nucleus basalis neurons demonstrated by in vitro pharmacological and immunohistochemical evidence in the guinea-pig brain. Eur J Neurosci 7:1502–1511

Fournier GN, Materi LM, Semba K, Rasmusson DD (2004) Cortical acetylcholine release and electroencephalogram activation evoked by ionotropic glutamate receptor agonists in the rat basal forebrain. Neuroscience 123:785–792

Fuller P, Sherman D, Pedersen NP, Saper CB, Lu J (2011) Reassessment of the structural basis of the ascending arousal system. J Comp Neurol 519:933–956

Gent TC, Bassetti C, Adamantidis AR (2018) Sleep-wake control and the thalamus. Curr Opin Neurobiol 52:188–197

Gerashchenko D, Salin-Pascual R, Shiromani PJ (2001) Effects of hypocretin-saporin injections into the medial septum on sleep and hippocampal theta. Brain Res 913:106–115

Gritti I, Henny P, Galloni F, Mainville L, Mariotti M, Jones BE (2006) Stereological estimates of the basal forebrain cell population in the rat, including neurons containing choline acetyltransferase, glutamic acid decarboxylase or phosphate-activated glutaminase and colocalizing vesicular glutamate transporters. Neuroscience 143:1051–1064

Gritti I, Manns ID, Mainville L, Jones BE (2003) Parvalbumin, calbindin, or calretinin in cortically projecting and GABAergic, cholinergic, or glutamatergic basal forebrain neurons of the rat. J Comp Neurol 458:11–31

Haas H, Panula P (2003) The role of histamine and the tuberomamillary nucleus in the nervous system. Nat Rev Neurosci 4:121–130

Hajszan T, Zaborszky L (2002) Direct catecholaminergic-cholinergic interactions in the basal forebrain III. Adrenergic innervation of choline acetyltransferase-containing neurons in the rat. J Comp Neurol 449:141–157

Hallanger AE, Levey AI, Lee HJ, Rye DB, Wainer BH (1987) The origins of cholinergic and other subcortical afferents to the thalamus in the rat. J Comp Neurol 262:105–124

Hallanger AE, Wainer BH (1988) Ascending projections from the pedunculopontine tegmental nucleus and the adjacent mesopontine tegmentum in the rat. J Comp Neurol 274:483–515

Hamamoto M, Kiyokage E, Sohn J, Hioki H, Harada T, Toida K (2017) Structural basis for cholinergic regulation of neural circuits in the mouse olfactory bulb. J Comp Neurol 525:574–591

Han Y, Shi YF, Xi W, Zhou R, Tan ZB, Wang H, Li XM, Chen Z, Feng GP, Luo MM, Huang ZL, Duan SM, Yu YQ (2014) Selective Activation of Cholinergic Basal Forebrain Neurons Induces Immediate Sleep-wake Transitions. Curr Biol 24:693–698

Hassani OK, Lee MG, Henny P, Jones BE (2009) Discharge profiles of identified GABAergic in comparison to cholinergic and putative glutamatergic basal forebrain neurons across the sleep-wake cycle. J Neurosci Off J Soc Neurosci 29:11828–11840

Henny P, Jones BE (2006) Vesicular glutamate (VGluT), GABA (VGAT), and acetylcholine (VAChT) transporters in basal forebrain axon terminals innervating the lateral hypothalamus. J Comp Neurol 496:453–467

Henny P, Jones BE (2008) Projections from basal forebrain to prefrontal cortex comprise cholinergic, GABAergic and glutamatergic inputs to pyramidal cells or interneurons. Eur J Neurosci 27:654–670

Hobson JA, McCarley RW, Wyzinski PW (1975) Sleep cycle oscillation: reciprocal discharge by two brainstem neuronal groups. Science 189:55–58

Hudetz AG, Wood JD, Kampine JP (2003) Cholinergic reversal of isoflurane anesthesia in rats as measured by cross-approximate entropy of the electroencephalogram. Anesthesiology 99:1125–1131

Hur EE, Edwards RH, Rommer E, Zaborszky L (2009) Vesicular glutamate transporter 1 and vesicular glutamate transporter 2 synapses on cholinergic neurons in the sublenticular gray of the rat basal forebrain: a double-label electron microscopic study. Neuroscience 164:1721–1731

Irmak SO, de Lecea L (2014) Basal forebrain cholinergic modulation of sleep transitions. Sleep 37:1941–1951

Jasper HH, Tessier J (1971) Acetylcholine Liberation From Cerebral Cortex during Paradoxical (Rem) Sleep. Science 172:601

JimenezCapdeville ME, Dykes RW, Myasnikov AA (1997) Differential control of cortical activity by the basal forebrain in rats: a role for both cholinergic and inhibitory influences. J Comp Neurol 381:53–67

Jones BE (2004) Activity, modulation and role of basal forebrain cholinergic neurons innervating the cerebral cortex. Prog Brain Res 145:157–169

Jones BE (2008) Modulation of cortical activation and behavioral arousal by cholinergic and orexinergic systems. Ann N Y Acad Sci 1129:26–34

Jones BE, Cuello AC (1989) Afferents to the basal forebrain cholinergic cell area from pontomesencephalic--catecholamine, serotonin, and acetylcholine—neurons. Neuroscience 31:37–61

Jones BE, Paré M, Beaudet A (1986) Retrograde labeling of neurons in the brain stem following injections of [3H]choline into the rat spinal cord. Neuroscience 18:901–916

Jourdain A, Semba K, Fibiger HC (1989) Basal forebrain and mesopontine tegmental projections to the reticular thalamic nucleus: an axonal collateralization and immunohistochemical study in the rat. Brain Res 505:55–65

Jouvet M (1972) The role of monoamines and acetylcholine-containing neurons in the regulation of the sleep-waking cycle. Ergeb Physiol Biol Chem Exp Pharmakol 64:166–307

Kaisti KK, Langsjo JW, Aalto S, Oikonen V, Sipila H, Teras M, Hinkka S, Metsahonkala L, Scheinin H (2003) Effects of sevoflurane, propofol and adjunct nitrous oxide on regional cerebral blood flow, oxygen consumption, and blood volume in humans. Anesthesiology 99:603–613

Kalinchuk AV, McCarley RW, Stenberg D, Porkka-Heiskanen T, Basheer R (2008) The role of cholinergic Basal forebrain neurons in adenosine-mediated homeostatic control of sleep: lessons from 192 Igg-Saporin lesions. Neuroscience 157:238–253

Kang DS, Ding MZ, Topchiy I, Kocsis B (2017) Reciprocal interactions between medial septum and hippocampus in theta generation: Granger causality decomposition of mixed spike-field recordings. Front Neuroanat 11:120

Kapp BS, Supple WF Jr, Whalen PJ (1994) Effects of electrical stimulation of the amygdaloid central nucleus on neocortical arousal in the rabbit. Behav Neurosci 108:81–93

Kaur S, Junek A, Black MA, Semba K (2008) Effects of ibotenate and 192IgG-Saporin lesions of the nucleus basalis magnocellularis/substantia innominata on spontaneous sleep and wake states and on recovery sleep after sleep deprivation in rats. J Neurosci 28:491–504

Kenny JD, Chemali JJ, Cotten JF, Van Dort CJ, Kim SE, Ba D, Taylor NE, Brown EN, Solt K (2016) Physostigmine and methylphenidate induce distinct arousal states during isoflurane general anesthesia in rats. Anesth Analg 123:1210–1219

Khateb A, Fort P, Alonso A, Jones BE, Muhlethaler M (1993) Pharmacological and immunohistochemical evidence for serotonergic modulation of cholinergic nucleus basalis neurons. Eur J Neurosci 5:541–547

Kia HK, Brisorgueil MJ, Daval G, Langlois X, Hamon M, Verge D (1996) Serotonin1A receptors are expressed by a subpopulation of cholinergic neurons in the rat medial septum and diagonal band of Broca--a double immunocytochemical study. Neuroscience 74:143–154

Kim JH, Jung AH, Jeong D, Choi I, Kim K, Shin S, Kim SJ, Lee SH (2016) Selectivity of neuromodulatory projections from the Basal forebrain and locus ceruleus to primary sensory cortices. J Neurosci 36:5314–5327

Kim T, Thankachan S, McKenna JT, McNally JM, Yang C, Choi JH, Chen L, Kocsis B, Deisseroth K, Strecker RE, Basheer R, Brown RE, McCarley RW (2015) Cortically projecting basal forebrain parvalbumin neurons regulate cortical gamma band oscillations. Proc Natl Acad Sci U S A 112:3535–3540

Kirby LG, Pernar L, Valentino RJ, Beck SG (2003) Distinguishing characteristics of serotonin and non-serotonin-containing cells in the dorsal raphe nucleus: electrophysiological and immunohistochemical studies. Neuroscience 116:669–683

Kirk IJ, Oddie SD, Konopacki J, Bland BH (1996) Evidence for differential control of posterior hypothalamic, supramammillary, and medial mammillary theta-related cellular discharge by ascending and descending pathways. J Neurosci 16:5547–5554

Kreindler A, Steriade M (1964) Eeg patterns of arousal and sleep induced by stimulating various amygdaloid levels in the cat. Arch Ital Biol 102:576–586

Kroeger D, Ferrari LL, Petit G, Mahoney CE, Fuller PM, Arrigoni E, Scammell TE (2017) Cholinergic, glutamatergic, and GABAergic neurons of the pedunculopontine tegmental nucleus have distinct effects on sleep/wake behavior in mice. J Neurosci 37:1352–1366

Laalou FZ, de Vasconcelos AP, Oberling P, Jeltsch H, Cassel JC, Pain L (2008) Involvement of the basal cholinergic forebrain in the mediation of general (Propofol) anesthesia. Anesthesiology 108:888–896

Lavoie B, Parent A (1994) Pedunculopontine nucleus in the squirrel-monkey—cholinergic and glutamatergic projections to the substantia-nigra. J Comp Neurol 344:232–241

Lee MG, Hassani OK, Alonso A, Jones BE (2005) Cholinergic basal forebrain neurons burst with theta during waking and paradoxical sleep. J Neurosci 25:4365–4369

Leranth C, Kiss J (1996) A population of supramammillary area calretinin neurons terminating on medial septal area cholinergic and lateral septal area calbindin-containing cells are aspartate/glutamatergic. J Neurosci 16:7699–7710

Leranth C, Vertes RP (1999) Median raphe serotonergic innervation of medial septum diagonal band of broca (MSDB) parvalbumin-containing neurons: possible involvement of the MSDB in the desynchronization of the hippocampal EEG. J Comp Neurol 410:586–598

Leung LS, Ma J, Shen B, Nachim I, Luo T (2013) Medial septal lesion enhances general anesthesia response. Exp Neurol 247:419–428

Leung LS, Petropoulos S, Shen B, Luo T, Herrick I, Rajakumar N, Ma J (2011) Lesion of cholinergic neurons in nucleus basalis enhances response to general anesthetics. Exp Neurol 228:259–269

Leung LS, Shen BX, Rajakumar N, Ma JY (2003) Cholinergic activity enhances hippocampal long-term potentiation in CA1 during walking in rats. J Neurosci 23:9297–9304

Lin JS, Anaclet C, Sergeeva OA, Haas HL (2011) The waking brain: an update. Cell Mol Life Sci 68:2499–2512

Lin JS, Sakai K, Vannimercier G, Jouvet M (1989) A critical role of the posterior hypothalamus in the mechanisms of wakefulness determined by microinjection of muscimol in freely moving cats. Brain Res 479:225–240

Lin SC, Brown RE, Hussain Shuler MG, Petersen CC, Kepecs A (2015) Optogenetic dissection of the basal forebrain neuromodulatory control of cortical activation, plasticity, and cognition. J Neurosci 35:13896–13903

Lopez-Hernandez GY, Ananth M, Jiang L, Ballinger EC, Talmage DA, Role LW (2017) Electrophysiological properties of basal forebrain cholinergic neurons identified by genetic and optogenetic tagging. J Neurochem 142:103–110

Losier BJ, Semba K (1993) Dual projections of single cholinergic and aminergic brainstem neurons to the thalamus and basal forebrain in the rat. Brain Res 604:41–52

Lu J, Sherman D, Devor M, Saper CB (2006) A putative flip-flop switch for control of REM sleep. Nature 441:589–594

Luo T, Leung LS (2009) Basal forebrain histaminergic transmission modulates electroencephalographic activity and emergence from isoflurane anesthesia. Anesthesiology 111:725–733

Luppi PH, Billwiller F, Fort P (2017) Selective activation of a few limbic structures during paradoxical (REM) sleep by the claustrum and the supramammillary nucleus: evidence and function. Curr Opin Neurobiol 44:59–64

Luppi PH, Clement O, Sapin E, Gervasoni D, Peyron C, Leger L, Salvert D, Fort P (2011) The neuronal network responsible for paradoxical sleep and its dysfunctions causing narcolepsy and rapid eye movement (REM) behavior disorder. Sleep Med Rev 15:153–163

Manns ID, Mainville L, Jones BE (2001) Evidence for glutamate, in addition to acetylcholine and GABA, neurotransmitter synthesis in basal forebrain neurons projecting to the entorhinal cortex. Neuroscience 107:249–263

Marazioti A, Kastellakis A, Antoniou K, Papasava D, Thermos K (2005) Somatostatin receptors in the ventral pallidum/substantia innominata modulate rat locomotor activity. Psychopharmacology 181:319–326

Martin-Cortecero J, Nunez A (2014) Tactile response adaptation to whisker stimulation in the lemniscal somatosensory pathway of rats. Brain Res 1591:27–37

Massimini M, Ferrarelli F, Huber R, Esser SK, Singh H, Tononi G (2005) Breakdown of cortical effective connectivity during sleep. Science 309:2228–2232

McDonald AJ (1991) Topographical organization of amygdaloid projections to the caudatoputamen, nucleus accumbens, and related striatal-like areas of the rat brain. Neuroscience 44:15–33

McGinty DJ, Harper RM (1976) Dorsal raphe neurons: depression of firing during sleep in cats. Brain Res 101:569–575

McKenna JT, Cordeira JW, Jeffrey BA, Ward CP, Winston S, McCarley RW, Strecker RE (2009) c-Fos protein expression is increased in cholinergic neurons of the rodent basal forebrain during spontaneous and induced wakefulness. Brain Res Bull 80:382–388

Miller JD, Farber J, Gatz P, Roffwarg H, German DC (1983) Activity of mesencephalic dopamine and non-dopamine neurons across stages of sleep and walking in the rat. Brain Res 273:133–141

Mochizuki T, Arrigoni E, Marcus JN, Clark EL, Yamamoto M, Honer M, Borroni E, Lowell BB, Elmquist JK, Scammell TE (2011) Orexin receptor 2 expression in the posterior hypothalamus rescues sleepiness in narcoleptic mice. Proc Natl Acad Sci U S A 108:4471–4476

Momiyama T (2010) Developmental increase in D1-like dopamine receptor-mediated inhibition of glutamatergic transmission through P/Q-type channel regulation in the basal forebrain of rats. Eur J Neurosci 32:579–590

Momiyama T, Nishijo T (2017) Dopamine and Serotonin-Induced Modulation of GABAergic and Glutamatergic Transmission in the Striatum and Basal Forebrain. Front Neuroanat 11:42

Monti JM (2010) The role of dorsal raphe nucleus serotonergic and non-serotonergic neurons, and of their

receptors, in regulating waking and rapid eye movement (REM) sleep. Sleep Med Rev 14:319–327
Monti JM, Jantos H (2015) The effects of systemic administration and local microinjection into the central nervous system of the selective serotonin 5-HT2C receptor agonist RO-600175 on sleep and wakefulness in the rat. Behav Pharmacol 26:418–426
Monti JM, Jantos H, Schechter LE (2013) The effects of systemic and local microinjection into the central nervous system of the selective serotonin 5-HT6 receptor agonist WAY-208466 on sleep and wakefulness in the rat. Behav Brain Res 249:65–74
Moore RY, Bloom FE (1978) Central catecholamine neuron systems- anatomy and physiology of the dopamine systems. Annu Rev Neurosci 1:129–169
Moruzzi G, Magoun HW (1949) Brain stem reticular formation and activation of the EEG. Electroencephalogr Clin Neurophysiol 1:455–473
Muzerelle A, Scotto-Lomassese S, Bernard JF, Soiza-Reilly M, Gaspar P (2016) Conditional anterograde tracing reveals distinct targeting of individual serotonin cell groups (B5-B9) to the forebrain and brainstem. Brain Struct Funct 221:535–561
Nadasdy Z, Varsanyi P, Zaborszky L (2010) Clustering of large cell populations: method and application to the basal forebrain cholinergic system. J Neurosci Methods 194:46–55
Nambu T, Sakurai T, Mizukami K, Hosoya Y, Yanagisawa M, Goto K (1999) Distribution of orexin neurons in the adult rat brain. Brain Res 827:243–260
Oddie SD, Stefanek W, Kirk IJ, Bland BH (1996) Intraseptal procaine abolishes hypothalamic stimulation-induced wheel-running and hippocampal theta field activity in rats. J Neurosci 16:1948–1956
Ongini E, Caporali MG, Massotti M (1985) Stimulation of dopamine D-1 receptors by SKF 38393 induces EEG desynchronization and behavioral arousal. Life Sci 37:2327–2333
Pace-Schott EF, Hobson JA (2002) The neurobiology of sleep: genetics, cellular physiology and subcortical networks, Nature reviews. Neuroscience 3:591–605
Pain L, Jeltsch H, Lehmann O, Lazarus C, Laalou FZ, Cassel JC (2000) Central cholinergic depletion induced by 192 IgG-saporin alleviates the sedative effects of propofol in rats. Brit J Anaesth 85:869–873
Paraskeva A, Papilas K, Fassoulaki A, Melemeni A, Papadopoulos G (2002) Physostigmine does not antagonize sevoflurane anesthesia assessed by bispectral index or enhances recovery. Anesth Analg 94:569–572
Pare D, Smith Y (1994) GABAergic projection from the intercalated cell masses of the amygdala to the basal forebrain in cats. J Comp Neurol 344:33–49
Pedersen NP, Ferrari L, Venner A, Wang JL, Abbott SBG, Vujovic N, Arrigoni E, Saper CB, Fuller PM (2017) Supramammillary glutamate neurons are a key node of the arousal system. Nat Commun 8:1405
Peyron C, Tighe DK, van den Pol AN, de Lecea L, Heller HC, Sutcliffe JG, Kilduff TS (1998) Neurons containing hypocretin (orexin) project to multiple neuronal systems. J Neurosci 18:9996–10015
Phillis JW (2005) Acetylcholine release from the central nervous system: a 50-year retrospective. Crit Rev Neurobiol 17:161–217
Pillay S, Vizuete JA, McCallum JB, Hudetz AG (2011) Norepinephrine infusion into nucleus basalis elicits microarousal in desflurane-anesthetized rats. Anesthesiology 115:733–742
Poulin AN, Guerci A, El Mestikawy S, Semba K (2006) Vesicular glutamate transporter 3 immunoreactivity is present in cholinergic basal forebrain neurons projecting to the basolateral amygdala in rat. J Comp Neurol 498:690–711
Ramesh V, Thakkar MM, Strecker RE, Basheer R, McCarley RW (2004) Wakefulness-inducing effects of histamine in the basal forebrain of freely moving rats. Behav Brain Res 152:271–278
Rasmusson DD, Clow K, Szerb JC (1994) Modification of neocortical acetylcholine-release and electroencephalogram desynchronization due to brain-stem stimulation by drugs applied to the Basal forebrain. Neuroscience 60:665–677
Renouard L, Billwiller F, Ogawa K, Clement O, Camargo N, Abdelkarim M, Gay N, Scote-Blachon C, Toure R, Libourel PA, Ravassard P, Salvert D, Peyron C, Claustrat B, Leger L, Salin P, Malleret G, Fort P, Luppi PH (2015) The supramammillary nucleus and the claustrum activate the cortex during REM sleep. Sci Adv 1:e1400177
Rho HJ, Kim JH, Lee SH (2018) Function of selective neuromodulatory projections in the mammalian cerebral cortex: comparison between cholinergic and noradrenergic systems. Front Neural Circuit 12:47
Rye DB, Wainer BH, Mesulam MM, Mufson EJ, Saper CB (1984) Cortical projections arising from the basal forebrain: a study of cholinergic and noncholinergic components employing combined retrograde tracing and immunohistochemical localization of choline acetyltransferase. Neuroscience 13:627–643
Sainsbury RS, Bland BH (1981) The effects of selective septal lesions on theta production in CA1 and the dentate gyrus of the hippocampus. Physiol Behav 26:1097–1101
Saito H, Sakai K, Jouvet M (1977) Discharge patterns of the nucleus parabrachialis lateralis neurons of the cat during sleep and waking. Brain Res 134:59–72
Salib M, Joshi A, Katona L, Howarth M, Micklem B, Somogyi P, Viney TJ (2019) GABAergic medial septal neurons with low-rhythmic firing innervating the dentate gyrus and hippocampal area CA3. J Neurosci 39 (23):4527–4549
Sallanon M, Denoyer M, Kitahama K, Aubert C, Gay N, Jouvet M (1989) Long-lasting insomnia induced by preoptic neuron lesions and its transient reversal by muscimol injection into the posterior hypothalamus in the cat. Neuroscience 32:669–683
Saper CB, Fuller PM, Pedersen NP, Lu J, Scammell TE (2010) Sleep state switching. Neuron 68:1023–1042

Sarter M, Lustig C, Howe WM, Gritton H, Berry AS (2014) Deterministic functions of cortical acetylcholine. Eur J Neurosci 39:1912–1920

Schwarz LA, Luo L (2015) Organization of the locus coeruleus-norepinephrine system. Curr Biol 25: R1051–R1056

Semba K (2004) Phylogenetic and ontogenetic aspects of the basal forebrain cholinergic neurons and their innervation of the cerebral cortex. Prog Brain Res 145:3–43

Semba K, Fibiger HC (1989) Organization of central cholinergic systems. Progr Brain Res 79:37–63

Semba K, Reiner PB, McGeer EG, Fibiger HC (1988) Brainstem afferents to the magnocellular basal forebrain studied by axonal transport, immunohistochemistry, and electrophysiology in the rat. J Comp Neurol 267:433–453

Servos P, Barke KE, Hough LB, Vanderwolf CH (1994) Histamine does not play an essential role in electrocortical activation during waking behavior. Brain Res 636:98–102

Shiromani PJ, Winston S, McCarley RW (1996) Pontine cholinergic neurons show Fos-like immunoreactivity associated with cholinergically induced REM sleep. Mol Brain Res 38:77–84

Sim JA, Griffith WH (1996) Muscarinic inhibition of glutamatergic transmissions onto rat magnocellular basal forebrain neurons in a thin-slice preparation. Eur J Neurosci 8:880–891

Smiley JF, Subramanian M, Mesulam MM (1999) Monoaminergic-cholinergic interactions in the primate basal forebrain. Neuroscience 93:817–829

Smith KS, Csordas A, Hajszan T, Zaborszky L (2001) Somatostatin-containing neurons in the basal forebrain receive input from the ventral tegmental mesencephalon. Soc Neurosci 27:599.15

Steinfels GF, Heym J, Strecker RE, Jacobs BL (1983) Behavioral correlates of dopaminergic unit activity in freely moving cats. Brain Res 258:217–228

Steriade M (1995) Neuromodulatory systems of thalamus and neocortex. Semin Neurosci 7:361–370

Strecker RE, Morairty S, Thakkar MM, Porkka-Heiskanen T, Basheer R, Dauphin LJ, Rainnie DG, Portas CM, Greene RW, McCarley RW (2000) Adenosinergic modulation of basal forebrain and preoptic/anterior hypothalamic neuronal activity in the control of behavioral state. Behav Brain Res 115:183–204

Sutcliffe JG, de Lecea L (2002) The hypocretins: setting the arousal threshold. Nat Rev Neurosci 3:339–349

Tai SK, Ma J, Leung LS (2014) Medial septal cholinergic neurons modulate isoflurane anesthesia. Anesthesiology 120:392–402

Thakkar MM (2011) Histamine in the regulation of wakefulness. Sleep Med Rev 15:65–74

Thakkar MM, Ramesh V, Strecker RE, McCarley RW (2001) Microdialysis perfusion of orexin-A in the basal forebrain increases wakefulness in freely behaving rats. Arch Ital Biol 139:313–328

Thannickal TC, Moore RY, Nienhuis R, Ramanathan L, Gulyani S, Aldrich M, Cornford M, Siegel JM (2000) Reduced number of hypocretin neurons in human narcolepsy. Neuron 27:469–474

Trulson ME, Preussler DW (1984) Dopamine-containing ventral tegmental area neurons in freely moving cats—activity during the sleep-waking cycle and effects of stress. Exp Neurol 83:367–377

Unal CT, Golowasch JP, Zaborszky L (2012) Adult mouse basal forebrain harbors two distinct cholinergic populations defined by their electrophysiology. Front Behav Neurosci 6:21

Van Dort CJ, Zachs DP, Kenny JD, Zheng S, Goldblum RR, Gelwan NA, Ramos DM, Nolan MA, Wang KR, Weng FJ, Lin YX, Wilson MA, Brown EN (2015) Optogenetic activation of cholinergic neurons in the PPT or LDT induces REM sleep. Proc Natl Acad Sci U S A 112:584–589

Vanderwolf CH, Baker GB (1986) Evidence that serotonin mediates non-cholinergic neocortical low-voltage fast activity, non-cholinergic hippocampal rhythmical slow activity and contributes to intelligent behavior. Brain Res 374:342–356

Vertes RP (2015) Major diencephalic inputs to the hippocampus: supramammillary nucleus and nucleus reuniens. Circuitry and function. Prog Brain Res 219:121–144

Vertes RP, Kocsis B (1997) Brainstem-diencephalo-septohippocampal systems controlling the theta rhythm of the hippocampus. Neuroscience 81:893–926

Wang YQ, Li R, Zhang MQ, Zhang Z, Qu WM, Huang ZL (2015) The neurobiological mechanisms and treatments of REM sleep disturbances in depression. Curr Neuropharmacol 13:543–553

Weber F, Dan Y (2016) Circuit-based interrogation of sleep control. Nature 538:51–59

Webster HH, Jones BE (1988) Neurotoxic lesions of the dorsolateral pontomesencephalic tegmentum cholinergic cell area in the cat. 2. effects upon sleep waking states. Brain Res 458:285–302

Wenk GL, Stoehr JD, Quintana G, Mobley S, Wiley RG (1994) Behavioral, biochemical, histological, and electrophysiological effects of 192-Igg-Saporin injections into the Basal forebrain of rats. J Neurosci 14:5986–5995

Wisor JP, Nishino S, Sora I, Uhl GH, Mignot E, Edgar DM (2001) Dopaminergic role in stimulant-induced wakefulness. J Neurosci 21:1787–1794

Wu M, Zaborszky L, Hajszan T, van den Pol AN, Alreja M (2004) Hypocretin/orexin innervation and excitation of identified septohippocampal cholinergic neurons. J Neurosci 24:3527–3536

Xu M, Chung S, Zhang S, Zhong P, Ma C, Chang WC, Weissbourd B, Sakai N, Luo L, Nishino S, Dan Y (2015) Basal forebrain circuit for sleep-wake control. Nat Neurosci 18:1641–1647

Yang C, McKenna JT, Zant JC, Winston S, Basheer R, Brown RE (2014) Cholinergic neurons excite

cortically projecting basal forebrain GABAergic neurons. J Neurosci 34:2832–2844
Yu X, Franks NP, Wisden W (2018) Sleep and sedative states induced by targeting the histamine and noradrenergic systems. Front Neural Circuits 12:4
Zaborszky L (1989) Afferent connections of the forebrain cholinergic projection neurons, with special reference to monoaminergic and peptidergic fibers. EXS 57:12–32
Zaborszky L, Cullinan WE (1989) Hypothalamic axons terminate on forebrain cholinergic neurons—an ultrastructural double-labeling study using Pha-L tracing and chat immunocytochemistry. Brain Res 479:177–184
Zaborszky L, Cullinan WE, Braun A (1991) Afferents to basal forebrain cholinergic projection neurons: an update. Adv Exp Med Biol 295:43–100
Zaborszky L, Cullinan WE, Luine VN (1993) Catecholaminergic-cholinergic interaction In the Basal forebrain. Progr Brain Res 98:31–49
Zaborszky L, Duque A (2000) Local synaptic connections of basal forebrain neurons. Behav Brain Res 115:143–158
Zaborszky L, Gaykema RP, Swanson DJ, Cullinan WE (1997) Cortical input to the basal forebrain. Neuroscience 79:1051–1078
Zaborszky L, Leranth C, Heimer L (1984) Ultrastructural evidence of amygdalofugal axons terminating on cholinergic cells of the rostral forebrain. Neurosci Lett 52:219–225
Zaborszky L, Luine VN (1987) Evidence for existence of monoaminergic-cholinergic interactions in the basal forebrain. J Cell Biochem:187–187
Zaborszky L, Pang K, Somogyi J, Nadasdy Z, Kallo I (1999) The basal forebrain corticopetal system revisited. Ann N Y Acad Sci 877:339–367
Zant JC, Kim T, Prokai L, Szarka S, McNally J, McKenna JT, Shukla C, Yang C, Kalinchuk AV, McCarley RW, Brown RE, Basheer R (2016) Cholinergic neurons in the Basal forebrain promote wakefulness by actions on neighboring non-cholinergic neurons: an opto-dialysis study. J Neurosci 36:2057–2067
Zant JC, Rozov S, Wigren HK, Panula P, Porkka-Heiskanen T (2012) Histamine release in the Basal forebrain mediates cortical activation through cholinergic neurons. J Neurosci 32:13244–13254
Zhang H, Lin SC, Nicolelis MA (2011) A distinctive subpopulation of medial septal slow-firing neurons promote hippocampal activation and theta oscillations. J Neurophysiol 106:2749–2763
Zhang LN, Li ZJ, Tong L, Guo C, Niu JY, Hou WG, Dong HL (2012) Orexin-A facilitates emergence from propofol anesthesia in the rat. Anesth Analg 115:789–796
Zhang LN, Yang C, Ouyang PR, Zhang ZC, Ran MZ, Tong L, Dong HL, Liu Y (2016) Orexin-A facilitates emergence of the rat from isoflurane anesthesia via mediation of the basal forebrain. Neuropeptides 58:7–14
Zheng Y, Feng S, Zhu X, Jiang W, Wen P, Ye F, Rao X, Jin S, He X, Xu F (2018) Different subgroups of cholinergic neurons in the basal forebrain are distinctly innervated by the olfactory regions and activated differentially in olfactory memory retrieval. Front Neural Circuits 12:99
Zucker J (1991) Central cholinergic depression reduces mac for isoflurane in rats. Anesth Analg 72:790–795

The manufacturer's authorised representative in the EU is Springer Nature Customer Service Centre GmbH, Europaplatz 3, 69115 Heidelberg, Germany. If you have any concerns regarding our products, please contact ProductSafety@springernature.com

Printed and bound by CPI Group (UK) Ltd, Croydon, CR0 4YY
30/09/2026
02225574-0020